Oak Ridge
National Laboratory

Oak Ridge National Laboratory
The First Fifty Years

Leland Johnson

Daniel Schaffer

The University of Tennessee Press / Knoxville

All illustrations courtesy of Oak Ridge National Laboratory.

Library of Congress Cataloging-in-Publication Data

Johnson, Leland, 1937–
    Oak Ridge National Laboratory: the first fifty years / Leland Johnson,
Daniel Schaffer.
        p.   cm.
    Includes bibliographical references and index.
    ISBN 0-87049-853-3 (cl.: alk. paper)
    ISBN 0-87049-854-1 (pbk.: alk. paper)
    1. Oak Ridge National Laboratory—History. 2. Research institutes—
Tennessee—Oak Ridge—History. 3. Physical laboratories—Tennessee—
Oak Ridge—History. I. Schaffer, Daniel, 1950–   . II. Title.
QC789.2.U620257   1994
621.042'0720768'73—dc20   94-7667

Contents

Illustrations

Foreword

In 1947, when the Atomic Energy Commission (AEC) inherited from the Manhattan District the two scientific children of the Chicago Metallurgical Laboratory—the facilities at Oak Ridge and Argonne—it decided to designate them *national laboratories*. No one really knew what a national laboratory was. In a general way, these institutions were supposed to explore the peaceful uses of nuclear fission. But in choosing to call them *national* rather than *atomic energy* laboratories, the commission displayed extraordinary foresight, or perhaps luck. An atomic energy laboratory, in principle, goes out of business when the problems of atomic energy are solved, are taken over by commercial enterprises, or are regarded (as at present) as unimportant. A national laboratory, by contrast, is more or less ensured immortality by virtue of its name. The designation *national* implies that no problem of national importance—whether in energy, environment, defense, industrial competitiveness, or basic science—is necessarily off limits.

In the fifty years since Oak Ridge National Laboratory (ORNL) was founded, it has become a full-fledged national sociotechnological institute. Its capabilities span the entire range of scientific disciplines, including the social sciences. It addresses an array of problems whose only common attribute is their significance both to the nation and to the world.

Who, for example, would have predicted in 1943 that ORNL in 1993 would be one of the world's most powerful environmental laboratories, equipped to address economic, climatological, ecological, and energy aspects of global climate change? Or who would have expected ORNL to emerge as one of the world's most powerful centers for the development of high-temperature materials? How did this metamorphosis take place? After all, ORNL was conceived by its founding genius, Eugene P. Wigner, as a major center for nuclear reactor development.

In 1947, the Atomic Energy Commission, following the advice of the General Advisory Committee, decided that a laboratory in the hills of Tennessee could never achieve scientific distinction. It therefore designated Argonne as the country's only center for reactor development. The outlook for ORNL's survival was bleak. Robert Oppenheimer and James Conant were doubtful that the laboratory could survive; and I. I. Rabi, another prominent member of the General Advisory Committee, tried to persuade the scientists of ORNL to move, en masse, to the newly formed Brookhaven National Laboratory near New York City. So, ever since the laboratory was founded, ORNL's survival has been an overriding concern.

But, in a sense, survival is the overriding concern of all organizations, profit or nonprofit. That the weapons laboratories during these fifty years have not had this worry has not saved them from confronting their survival now that peace has broken out. The question, therefore, is not "Is survival your mission?" but "Have you accomplished *great things* that transcend the obvious, and ever-present, issue of survival?"

To record ORNL's transition from wartime pilot plant to national sociotechnical institute and to interpret its many achievements that transcend mere survival is the task well accomplished by historians Leland Johnson and Daniel Schaffer in *Oak Ridge National Laboratory: The First Fifty Years.*

"Gray eagles" such as myself who were present at the creation of the laboratory are falling off, one by one. With each of our deaths, another bit of organizational memory disappears. Yet this memory is an important element of organizational morale. Knowing and understanding how the laboratory overcame challenges to its very existence, and how it eventually achieved greatness, should serve to inspire the new generation of ORNL employees and help inform the public at large. For this accomplishment, the new generations, as well as the gray eagles, must be grateful to the authors of this splendid history.

Alvin W. Weinberg
Director (1955–73)
Oak Ridge National Laboratory

Preface

One of the world's premier scientific research centers, Oak Ridge National Laboratory represents a marriage between science and industrial technology forged for national defense during the throes of global war. Operated by Martin Marietta Energy Systems, it is the oldest national laboratory on its original site, site of the world's oldest nuclear reactor, and home to the Department of Energy's largest and most diversified multiprogram laboratory.

As a government-sponsored institution operated by a private corporation to advance science and technology in partnership with universities and industries, Oak Ridge, as well as other national laboratories, embodied a new approach to scientific and governmental administration. Because solutions to energy and environmental problems have been found as much in engineering and applied technology as in basic science, ORNL, since its inception, has offered a vital link between the two and has always carried an avowedly semi-industrial appearance clothed by an academic predisposition.

Celebrating fifty years of service to the United States in 1993, ORNL has changed the history of the nation and the world. As a remarkable and sometimes bewildering complex of sophisticated industrial, science, and educational activities in an isolated rural setting, ORNL encapsulates the ever-changing nature of the U.S. research agenda, reflecting on a small, institutional scale sweeping shifts in national and global concerns during the past fifty years.

In its early years, the laboratory employed 1,500 scientists and support staff housed in primitive wooden frame buildings. There people worked—often unknowingly—on the construction of a graphite reactor and the extraction of plutonium from uranium.

Since then, the laboratory has experienced many transitions. In the postwar years, it survived budget and staff retrenchments by focusing on nuclear science and the development of nuclear energy for peaceful uses. In the 1960s, it became the first national laboratory to turn to research tied only tangentially to nuclear energy. During the 1970s, it expanded its research agenda, in accord with shifting national priorities, to encompass all forms of energy and their impact on the environment. In the 1980s, it became a multiprogram laboratory of the Department of Energy (DOE), leading broad research initiatives responsive to national needs. By its fiftieth anniversary, ORNL had emerged as a premier global research center for issues related to energy, environment, and basic science and technology.

Currently employing about 4,500 people, the laboratory has a research agenda ranging from global warming to energy conservation to superconductivity to ozone-safe substitutes for chlorofluorocarbons. It is committed to improving national science education and to speeding the transfer of its technological developments to the commercial marketplace.

Since 1943, scientists and technicians at ORNL have confronted issues vital to human life and its environment. Established to create nuclear weapons of unprecedented destructive power, the supreme paradox of the laboratory's history is its subsequent contributions to energy, environment, health, and the economy. Today, millions of people each year benefit from the results of its research and development activities. Oak Ridge National Laboratory, in short, has a history worth noting and a future worth watching.

Alvin Trivelpiece
Director
Oak Ridge National Laboratory

Introduction

This history of the first fifty years of Oak Ridge National Laboratory, sponsored by ORNL, was prepared to commemorate its golden anniversary in 1993. The laboratory's Fiftieth-Year Celebration Committee provided direction and resources for the study, and we are grateful to its members for their guidance and encouragement. Don Trauger chaired the committee—composed of Ed Aebischer, William Alexander, Darryl Armstrong, Stanley Auerbach, Deborah Barnes, Waldo Cohn, Charles Coutant, Joanne Gailar, Carolyn Krause, Charles Kuykendall, Ellison Taylor, Mike Wilkinson, and Alex Zucker—all current or retired ORNL employees. Anne Calhoun and Kim Pepper, also staff members, coordinated the committee's work.

Our exploration of historical sources was facilitated by librarians Mary Alexander, Gabrielle Boudreaux, Deborah Cole, Robert Conrad, Nancy Gray, Dianne Griffith, Kendra Jones, William Myers, Vicki Punsalan, and Deborah York; by Linda Cabage, Ray Evans, and Lynn Rodems of Public Affairs; by Becky Lawson, Lowell Langford, Linda Crews, Shirlene Rudder, and Marie Swenson of Laboratory Records; by Carolyn Krause, James Pearce, and William Cabage of the Publications Division; and by Frank Hoffman and Anna Conover of Analysas Corporation. The authors appreciate their kind assistance.

For making available the resources of the Children's Museum of Oak Ridge, we owe special thanks to Jane Alderfer, James Overholt, and Selma Shapiro. Research assistants Susan Schexnayder, Cathy Shires, and Edythe Quinn provided invaluable insights into the voluminous materials, and administrative assistant Becky Robinson helped keep the information in order once it was collected. Marilyn Morgan, a graduate student in the University of Tennessee's Department of English, gave the manuscript a thorough and careful reading during its final draft stages.

For enlightenment and inspiring ideas, we are indebted to Laura Fermi, Richard Fox, Milton Lietzke, Herbert MacPherson, Herbert Pomerance, Herman Postma, Raymond Stoughton, Chet Thornton, Elaine Trauger, Alvin Trivelpiece, Alvin Weinberg, and a host of ORNL personnel who took time from their busy schedules for both formal interviews and informal chats that broadened our understanding of the laboratory's past.

❈

Astrophysicists tell us the space-time continuum and the behavior of light prevent us from seeing a true image of the present. Like it or not, these physicists say, only the past provides a clear portrait of our lives and behavior—a conclusion that historians are more than eager to share.

Unlike physicists and other scientists, however, historians and writers live in a world of changing human perceptions and behavior, not in a world of immutable natural laws and fixed physical phenomena. For these reasons, what follows should be considered *a* history, not *the* history, of ORNL. Except for rare instances (for example, the day the graphite reactor went critical), people will disagree about the relative importance of specific ORNL accomplishments and the relative contributions of various staff members. Problems of assessment and attribution, moreover, are compounded by problems of space, time, and memory.

For the writers, space limitations required selecting for discussion only a few of ORNL's many significant achievements, projects, and programs. For readers, fifty years of history dims memories and may place at odds what actually happened from what participants now think happened.

Despite these inevitable limitations, we hope this presentation of Oak Ridge National Laboratory's past will be conducive to a better understanding of its present, serving both as a guidepost for ORNL's strengths and weaknesses and as a road map for its future endeavors. We also hope that readers, through these pages, are able to share some of the joy, excitement, obstacles, and pride that have accompanied the staff's journey of discovery.

Leland Johnson
Daniel Schaffer
January 1994

Chapter 1

Wartime Laboratory

Spreading along broad valleys cut by the Clinch River and framed by the foothills of the Appalachian Mountains, Oak Ridge—in East Tennessee—is a pleasant place. Along its highest ridges, a person can gaze at the majestic, cloud-capped Great Smoky Mountains to the east and the stately, tree-covered Cumberland Plateau to the west.

Southern Appalachia, as a whole, is a region of unique character with folkways as rich as those of New York City's East Side or Louisiana's Cajun country. Its rugged, yet beautiful, terrain has proved fertile ground for a life-style defined by independence and self-determination.

At the time of the Japanese attack on Pearl Harbor on December 7, 1941, century-old family farms and small crossroads communities such as Scarborough, Wheat, Robertsville, and Elza occupied what was about to become the Oak Ridge reservation. Outsiders thought the region quaint, a throwback to the nineteenth-century frontier that time and progress had passed by.

In truth, the area experienced enormous change during the early twentieth century. On the upside, it felt the effects of Henry Ford's automobile and shared, to some extent, the comforts afforded by electricity; on the downside, it reeled from the aftershocks of the Great Depression that had rocked the economy and had exerted additional pressures on the region's fragile natural resources.

Located just twenty-five miles from the Tennessee Valley Authority's (TVA) corporate headquarters at Knoxville and just a few miles below TVA's huge Norris Dam on the Clinch River, the area was, in fact, a focal point of one of the nation's boldest experiments in social and economic

Clinton Laboratories under construction at the X-10 site in October 1943.

engineering. The tiny Wheat community, for example, had been selected for a TVA-inspired venture in cooperative agriculture.

Residents of the Oak Ridge area in 1941 did not feel bypassed by
history. But the advent of the automobile, the introduction of electricity, the hardships of the Great Depression, and direct participation in
an unprecedented government-sponsored social experiment did not
prepare them for what was about to happen.

· In early 1942, the Army Corps of Engineers identified a fifty-nine-
thousand-acre swatch of land between Black Oak Ridge to the north
and the Clinch River to the south as a federal reserve to serve as one of
three sites nationwide for the development of the atomic bomb. Residents received court orders to vacate their ancestral homes within weeks.
Thousands of scientists, engineers, and workers soon swarmed into Oak
Ridge to build and operate three huge facilities that would change the
history of the region and the world forever.

On the reservation's western edge rose K-25, or the gaseous diffusion
plant, a warehouselike structure covering more area than any building

East Tennessee's rural life would be changed forever by construction of the Manhattan Project.

ever built. Completed at a cost of $500 million and operated by twelve thousand workers, the K-25 plant separated uranium-235, an isotope suited for achieving nuclear fission, from uranium-238. On its northern edge, near the workers' city named Oak Ridge, rose the Y-12 plant where an electromagnetic method was used to separate uranium-235. Built for $427 million, the Y-12 plant employed twenty-two thousand workers. Near the reservation's southwest corner, about ten miles from Y-12, was the third plant, X-10.

Built between February and November 1943 for $12 million and employing only 1,513 people during the war, X-10 was much smaller than K-25 and Y-12. As a pilot plant for the larger plutonium plant built at Hanford, Washington, X-10 used neutrons emitted in the fission of uranium-235 to convert uranium-238 into a new element, plutonium-239. During the war, X-10 was called Clinton Laboratories, named after the nearby county seat of rural Anderson County; in 1948, Clinton Laboratories became Oak Ridge National Laboratory.

ORNL, which celebrated its fiftieth anniversary in 1993, has evolved

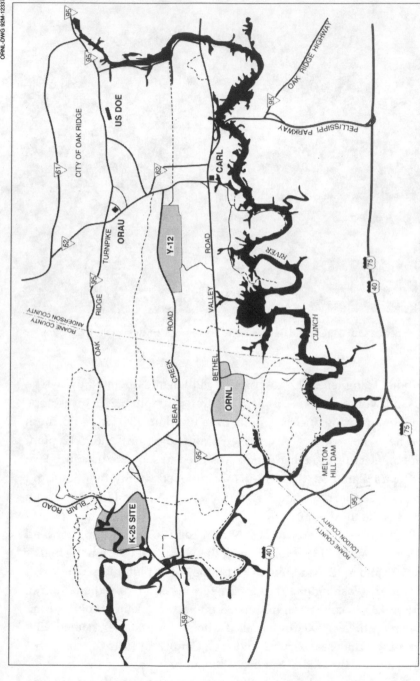

Map of the Oak Ridge vicinity showing the K-25, Y-12, and X-10 (ORNL) sites.

from a war-emergency pilot plant operated under the cloak of secrecy into one of the nation's outstanding research centers for energy, environment, basic research, and technology development. It currently employs about 4,500 people, including many scientists recognized internationally as experts in their fields. Laboratory undertakings range from studies of nuclear chemistry and physics to inquiries into global warming, energy conservation, superconductivity, and new materials. Its institutional roots, however, lie with the awesome power released by the splitting of atoms.

The laboratory's nuclear roots run deep and nourish much of its research, which is designed to improve the safety of commercial nuclear power, identify effective methods of managing nuclear waste, and achieve practical fusion power. The roots are not only deep, they are broadly international.

Supreme irony marks the laboratory's history. The institution, born during war, was propelled by a sense of urgency: If Hitler's scientists were to unleash atomic power first, Nazi Germany might place the entire world under a fascist fist. Yet the laboratory's present scientific excellence could not have been achieved without the camaraderie and sense of collective purpose that drive international science. Created to build a weapon capable of unprecedented destruction, ORNL became an institution that nurtures the ability of people to understand and transform their universe for the better. For this reason and more, its history merits the telling.

Laboratory Roots

The history of ORNL begins in three distinctly different places: Albert Einstein's retreat on Long Island, New York; the executive offices of the White House in Washington, D.C.; and university laboratories throughout the nation and overseas, especially at the University of Chicago. At its highest level, the scientific community is international in scope. As fascist dictators seized power in Europe during the 1930s, some of Europe's most renowned scientists fled to join colleagues in Britain and America. Among them were the German Albert Einstein, the Italian Enrico Fermi, and Hungarians Edward Teller, Leo Szilard, John von Neumann, and

Eugene Wigner. These brilliant minds enlisted in cooperative international efforts to develop atomic weapons and, later, nuclear energy. In the process, they significantly influenced twentieth-century history in general and the history of ORNL in particular. Eugene Wigner, in fact, has been called the "patron saint" of the laboratory.

Eugene Wigner, a pioneering chemical engineer and physicist from Budapest, Hungary, may have been the least known of the immigrant scientists. Completing a chemical engineering degree in Berlin in 1925, Wigner took a job at a Budapest tannery where his father also worked. Physics was his evening and weekend hobby. His friend John von Neumann called his attention to mathematical group theory, and Wigner soon published a series of technical papers that applied symmetry principles to problems of quantum mechanics. After two years at the tannery, he accepted an assistantship in theoretical physics in Berlin at the princely salary of thirty-two dollars per month.

In Berlin, Wigner established an international reputation as a physicist and, in 1930, Princeton University hired both him and von Neumann, each on a half-time basis. For a few years, the two friends commuted every six months between Berlin and Princeton until the Nazi government terminated their employment.

Wigner then went to the University of Wisconsin to work. There he devised a fundamental formula that enabled scientists to understand a neutron's energy variations when channeled through materials having different absorption capabilities for subatomic particles. At Wisconsin, he also discovered a university life that reached beyond academic circles to plain people who grew potatoes and milked cows, and he met scientists who repaired their cars and made home improvements. He later said that at Wisconsin he came to love his adopted country.

Returning to Princeton, Wigner studied solid-state physics and supervised the work of graduate students. His first student, Frederick Seitz, later became president of the National Academy of Sciences and of Rockefeller University; his second, John Bardeen, developed the transistor and twice received the Nobel Prize in physics.

The increasing strength of fascist governments in Europe deeply troubled Wigner. As a youngster, he had seen Hungary's enfeebled monarchy supplanted by brutal communist and then fascist governments. From personal experience, he developed an implacable enmity

Four scientists who advanced our understanding of nuclear physics: *left to right*, Walter Zinn, Leo Szilard, Eugene Wigner, and Alvin Weinberg.

toward totalitarian regimes. When he learned in early 1939 that two German chemists had discovered nuclear fission in uranium, Wigner recognized that this discovery could lead to both weapons of mass destruction and abundant energy for mass consumption.

Fearing Nazi Germany would initiate a crash program to develop atomic weapons, Wigner urged the U.S. government to support research on nuclear fission. He found an ally in his fellow countryman Leo Szilard, who in Hungary had attended the same schools as Wigner before emigrating to the United States.

Studying nuclear fission with Enrico Fermi at Columbia University in New York City, Szilard needed additional funds to continue his experiments with uranium and graphite. Wigner gladly lent his support to Szilard's efforts. Because other scientists were lobbying authorities with their own weapon schemes, Wigner and Szilard found their campaign for nuclear fission research moved so slowly they seemed to be "swimming in syrup."

Thinking that federal officials in Washington, D.C., would be more likely to listen to the famous Albert Einstein, an old acquaintance from Berlin, Wigner and Szilard sought him out in July 1939. Learning he had left Princeton, New Jersey, to vacation on Long Island, they drove there, located Einstein's cabin, and explained to him why the United States should initiate fission research before German scientists developed an atomic weapon. As Wigner later recalled:

> Einstein understood it in half a minute. It was really uncanny how he dictated a letter in German with enormous readiness. It is not easy to formulate and phrase things at once in a printable manner. He did. I translated that into English. Szilard and Teller went out, and Einstein signed it. Alexander Sachs took it to Washington. This helped greatly in initiating the uranium project.

In October 1939, President Franklin Roosevelt appointed a committee of prominent scientists and government administrators to manage federally funded scientific research. Wigner, Szilard, and Edward Teller met with the committee and requested six thousand dollars to purchase graphite for fission experiments. They listened to an army officer on the committee expound at length upon his theory that civilian and troop morale, not experimental weapons, won wars.

Szilard later recalled that "suddenly Wigner, the most polite of us, interrupted him. He said in his high-pitched voice that it was very interesting for him to hear this, and if this is correct, perhaps one should take a second look at the budget of the Army, and maybe the budget should be cut." The officer glared in silence at Wigner, and the committee agreed to provide the money for the experiments.

This first six thousand dollars of federal funding for nuclear energy research launched a vast, multibillion-dollar program that has continued unabated under the successive management of the U.S. Army, the Atomic Energy Commission (AEC), the Energy Research and Development Administration (ERDA), and the Department of Energy (DOE). The program has had direct and lasting ties to atomic research, development, and production sites across the United States, including Oak Ridge.

Initial funds for the uranium and graphite experiments, however, were not released until late 1940. Wigner became increasingly exasperated as the irreplaceable months passed. After the war, he contended that the delay, largely due to bureaucratic foot-dragging, cost many lives and billions of dollars.

American scientists, nevertheless, made vital advances in the interim. At Columbia University, in March 1940, John Dunning and his colleagues demonstrated that fission occurred more readily in the isotope uranium-235 than in uranium-238, but only 1 of 140 uranium atoms was the 235 isotope. Using cyclotrons at the University of California, in 1940 Edwin McMillan and Philip Abelson discovered the first transuranium element, number 93 on the atomic periodic table. They named it neptunium. A year later, Glenn Seaborg and colleagues discovered element 94 (the decay product of the newly synthesized number 93), naming it plutonium (in the planetary sequence Uranus, Neptune, Pluto), and demonstrated its fissionability.

Two doors to atomic weapons and energy thus had been opened for future exploration: less-common uranium-235 could be separated from more-common uranium-238 for weapons use, and uranium-238 could be bombarded with neutrons—in a nuclear pile or reactor—to produce plutonium that could then be chemically extracted for weapons production.

Metallurgical Laboratory

The day after the Japanese attacked Pearl Harbor on December 7, 1941, Arthur Compton, a Nobel laureate at the University of Chicago, contacted Eugene Wigner to discuss the possibility of consolidating nationwide atomic research efforts in Chicago. At meetings in January 1942, Compton brought together scientists experimenting with nuclear chain reactions at Princeton and Columbia universities, on the East Coast, with those investigating plutonium chemistry at the University of California to outline the objectives of this top-secret project.

Compton's schedule called for determining the feasibility of a nuclear chain reaction by July 1942, achieving the first self-sustaining chain reaction by January 1943, extracting the first plutonium from irradiated uranium-238 by January 1944, and producing the first atomic bomb by January 1945. In the end, all these deadlines were met except the last, which occurred six months later than planned.

Arthur Compton and Richard Doan headed the Metallurgical Laboratory at the University of Chicago. Doan, *on the right*, became research director of Clinton Laboratories in 1943.

To accomplish these objectives, Compton formed a "Metallurgical Laboratory" as cover at the University of Chicago and brought scientists from across the nation to this central location. The scientists' tasks were threefold: to develop chain-reacting "piles" for plutonium production, to devise methods for extracting plutonium from the irradiated uranium, and to design a weapon. Remaining in charge of the overall project, Compton selected Richard Doan as Metallurgical Laboratory director. An Indiana native, Doan had earned a physics degree from the University of Chicago in 1926 and had been a researcher for Western Electric and Phillips Petroleum before the war.

Compton also placed Glenn Seaborg in charge of the research on plutonium chemistry and assigned him the task of devising methods to separate plutonium from irradiated uranium in quantities sufficient for bomb production. To coordinate the theoretical and experimental phases of research associated with a chain reaction, Compton chose Eugene Wigner, Enrico Fermi, and Samuel Allison. Fermi continued his experiments with ever-larger piles of uranium and graphite, while Samuel

Allison directed a cyclotron group, including Canadian Arthur Snell, which assessed nuclear activities in uranium and graphite piles. Wigner and Snell later joined the X-10 staff.

Eugene Wigner headed the theoretical physics group crowded into the garrets of Eckart Hall on the University of Chicago campus. His "brain trust" of twenty scientists studied the arrangement, or lattice, of uranium and control materials for achieving a chain reaction and planned the design of nuclear reactors. Among Wigner's group were Gale Young, Kay Way, and Alvin Weinberg, all of whom later moved to Oak Ridge.

Having a chemical engineering background, Wigner also offered advice to Glenn Seaborg and his staff of University of California chemists, who were seeking to separate minuscule traces of plutonium from uranium irradiated in cyclotrons. This task was particularly challenging because to that point no one had isolated even a visible speck of plutonium. By September 1942, the team had obtained a few micrograms of plutonium for experimentation, but they needed much more for additional analysis.

In 1942, Compton brought Martin Whitaker, a North Carolinian who chaired New York University's Department of Physics, to Chicago to help Enrico Fermi and Walter Zinn build subcritical uranium and graphite piles. He later put Whitaker in charge of a laboratory under construction in the Argonne forest preserve on Chicago's southwest side. It was here that Compton initially planned to bring the first nuclear pile to critical mass. A strike by construction workers, however, prevented the laboratory's timely completion. As a result, Compton and Fermi decided to build a graphite pile housed in a squash court under the stands of the University of Chicago's stadium.

Leo Szilard and later Norman Hilberry were placed in charge of supplying materials for the pile experiments. They obtained impurity-free graphite from the National Carbon Company in Cleveland, Ohio, and the purest uranium metal available from Frank Spedding's research team at Ames, Iowa. George Boyd and chemists at Chicago analyzed the materials to assure the absence of impurities that might interfere with a nuclear reaction. Fermi and his colleagues then put the materials into a series of subcritical uranium and graphite piles built in what was to become the world's most famous squash court.

Fermi called them piles because, as the name implies, they were

$$-\frac{\hbar}{i}\frac{\partial}{\partial \tau} = \frac{p^2}{2m} \cdot \frac{Ze^2}{\tau}$$

$$\alpha = \frac{\hbar^2}{ec}$$

Enrico Fermi directed design of the first nuclear reactor at Chicago in 1942 and the graphite reactor at Oak Ridge in 1943.

stacks or piles of graphite blocks with lumps of uranium interspersed between them in specific lattice arrangements. Uranium formed the "core," or source of neutrons, and graphite served as a "moderator," slowing the neutrons to facilitate nuclear fission.

In truth, the piles were small, subcritical nuclear reactors cooled by air, but the name *reactor* did not supplant *pile* until 1952. Fermi gradually built larger subcritical piles, carefully measuring and recording neutron activity within them, edging toward the point where the pile would reach "critical mass" and the reaction would be self-sustaining.

On December 2, 1942, Fermi, Whitaker, and Zinn piled tons of graphite and uranium on the squash court to demonstrate a controlled nuclear reaction for visiting dignitaries standing on a balcony. Controlling the reaction with a rod coated with cadmium, a neutron-absorbing material, Fermi directed the phased withdrawal of the rod, carefully measuring the increased neutron flux within the pile. The pile went "critical," achieving self-sustaining status at 3:20 P.M., an event later hailed as the dawn of the atomic age.

Kermit Campbell holds a
container of uranium-238
shipped from Chicago to
Oak Ridge in 1943.

Having no shield to prevent a release of radiation, Fermi briefly operated this Chicago Pile 1, disassembled it, and in 1943 rebuilt it with concrete, radiation-protecting shielding as Chicago Pile 2 at the Argonne Laboratory.

Richard Fox, who rigged the control rod mechanism for Fermi's pile, stood behind Fermi on that fateful December afternoon fretting over the outcome of the first critical experiment. "The manual speed control was nothing more elaborate than a variable resistor," Fox recalled, "with a piece of cotton clothesline over a pulley and two lead weights to make it 'fail safe' and return to its zero position when released." Once the experiment succeeded and his concern that the clothesline would slip off the pulley proved unfounded, Fox recalled his elation: "It was as though we had discovered fire!"

After the dignitaries departed, Wigner brought out a bottle of Italian Chianti in honor of Fermi's achievement and shared toasts with the workers. He had carried the bottle from Princeton and later claimed it had taken more foresight to anticipate that the war would make Chi-

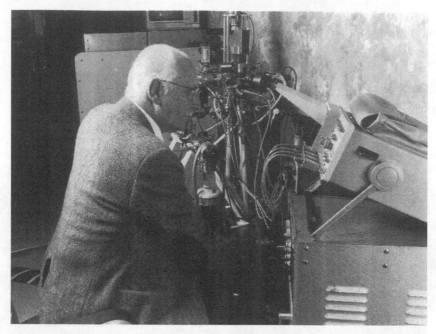

Richard Fox transferred from Chicago to Oak Ridge in 1943 and still worked at the laboratory in 1992.

anti a rare wine than that Fermi's chain reaction would succeed. Among the celebrants were Richard Fox and Ernest Wollan, who had monitored and recorded the radiation emitted by the reaction. Both left Chicago for Oak Ridge in 1943, where Wollan conducted neutron diffraction experiments and Fox applied his talents in the Instrumentation and Controls Division.

Producing sufficient quantities of plutonium for weapons use necessitated construction of large reactors that operated at high power levels, thus releasing a great deal of heat and radiation. To meet this goal, Metallurgical Laboratory engineers Thomas Moore and Miles Leverett, both recruited from the Humble Oil Company, began an intensive investigation of potentially larger reactor designs.

Scaling up Fermi's pile would not do, because extracting plutonium from the uranium would require tearing the pile apart each time and then reassembling it—a risky, time-consuming exercise. Moore and

Leverett developed a new design that used helium gas under pressure as the coolant to remove heat from the pile during a nuclear reaction. To extract the uranium without disassembling the graphite moderator, they designed holes or channels that extended through the graphite to allow the insertion of uranium rods. The rods could then be removed after they had been spent and irradiated.

Scientists agreed that thick shells of concrete would contain the radiation within the reactors, but they disagreed about methods for removing the heat. Fermi wanted an air-cooled reactor, with fans forcing air through channels alongside the uranium rods. Moore and Leverett preferred using helium gas under pressure. Szilard favored a liquid bismuth metal coolant, similar to the system he and Einstein had patented for refrigerators. And Wigner preferred plain river water, with uranium rods encased in aluminum to protect against water corrosion.

Wigner's water-cooling plan eventually was adopted for use in large reactors, but not before the decision to build Fermi's air-cooled graphite and uranium pilot reactor at Oak Ridge had been made.

The proposed pilot reactor would serve two purposes: First, it would test control-and-operation procedures; second, it would provide the larger quantities of plutonium required by the project's chemists. In mid-1942, Glenn Seaborg's chemical research group had used a lanthanum fluoride carrier process to separate micrograms of plutonium from uranium irradiated in cyclotrons; they now sought a means to achieve the separation on an industrial scale.

Among the various methods investigated for separating plutonium were the ion exchange and solvent extraction processes. Although not adopted in 1943, these studies provided the basis for the postwar separation of radioisotopes and the widely used solvent-extraction methods for recovering uranium and plutonium from spent nuclear fuel. In 1943, Seaborg and Du Pont chemist Charles Cooper settled on a small pilot plant using the lanthanum fluoride carrier built on the Chicago campus and another pilot plant using a bismuth phosphate carrier planned for Oak Ridge. In both cases, the separation would have to be conducted by remote control in "hot cells" encased in thick concrete to protect the chemists from radiation.

To the Hills

As the Metallurgical Laboratory's research continued, the government began to inspect potential sites for the planned industrial-scale uranium separation plants and pilot plutonium production and separation facilities. Officials desired an isolated inland site with plenty of water and abundant electric power.

At the recommendation of the War Production Board, Compton's chief of engineering, Thomas Moore, and two consulting engineers visited East Tennessee in April 1942. They found a desirable site bordering the Clinch River between the small towns of Clinton and Kingston that was served by two railroads and TVA electric power. Arthur Compton then inspected the site, approved it, and visited David Lilienthal, chairman of TVA, to describe the unfolding plans to purchase the land.

Lilienthal was dismayed by the news. He objected that the proposed site included land selected for an agricultural improvement program and suggested instead that Compton choose a site in western Kentucky near Paducah.

Compton refused to consider Lilienthal's alternative proposal and advised him that the land in East Tennessee would be taken through court action for immediate use. He urged Lilienthal not to question his judgment or inquire into the reasons for the purchase. "It was a bad precedent," Lilienthal later complained. "That particular site was not essential; another involving far less disruption in people's lives would have served as well, but arbitrary bureaucracy, made doubly powerful by military secrecy, had its way."

In June 1942, President Roosevelt assigned the army responsibility for managing uranium and plutonium plant construction and nuclear weapons production. High-ranking army officials, in turn, delegated this duty to Colonel James Marshall, commander of the Manhattan Engineer District headquartered initially in New York City and later relocated to Oak Ridge.

Because Fermi had not yet achieved a self-sustaining chain reaction, Marshall and army authorities postponed their efforts to acquire the land. The delay disturbed some scientists anxious not to lose ground to the Ger-

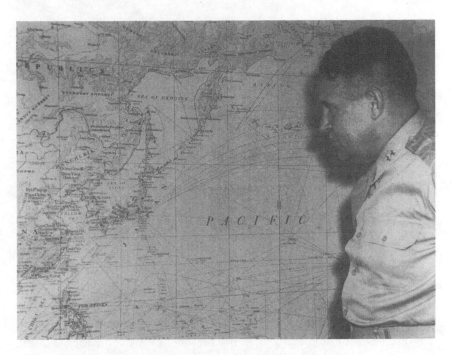

General Leslie R. Groves commanded the Manhattan Project.

mans. It also perturbed the hard-driving deputy chief of the U.S. Army Corps of Engineers, Brigadier General Leslie Groves.

Given command of the Manhattan Project in September 1942, Groves ordered immediate purchase of the reservation, first given the code name Kingston Demolition Range after the town south of the reservation and later renamed Clinton Engineer Works after the town to the north. The army sent an affable Kentuckian, Fred Morgan, to open a real estate office near the site and purchase the land through court condemnation, thereby securing clear title for its immediate use. About one thousand families on the reservation were paid for their land and forced to relocate. Existing structures were demolished or converted to other, war-related uses. New Bethel Valley Baptist Church at the X-10 site, for example, was used for storage.

To speed production of weapons materials, Groves selected experienced industrial contractors to build and operate the plants. In January 1943, he persuaded Du Pont to initiate construction of both the pilot facilities at X-10 in Oak Ridge and the full-scale reactors to be built later in Hanford, Washington. Involved in too many military projects and reluctant to undertake the work at X-10, Du Pont executives were persuaded to accept Grove's request partly through appeals to their patriotism. The contract stipulated that Du Pont would withdraw from the job at war's end, accept no work-related patents, and receive no payment other than their costs plus a one-dollar profit. After the war, Groves reported with amusement that government auditors allowed Du Pont a profit of only sixty-six cents because the company had finished its job ahead of schedule.

Groves called on the University of Chicago to operate the pilot plutonium plant planned at X-10, but scientists at the Metallurgical Laboratory in Chicago expressed initial dissatisfaction with this proposal. Wigner and others had wanted to design and construct the plants, and they were not interested in operating them after Du Pont had been given the jobs they had sought. Also, university scientists and administrators preferred building the pilot plant in the Argonne forest convenient to Chicago; the prospect of operating industrial facilities five hundred miles from their campus in the remote hills of Tennessee did not elicit much enthusiasm.

Groves and the army again used appeals to patriotism to help persuade the university to accept the challenge. The compromise called for Chicago to supply the managers and scientists needed for the operations and for Du Pont to mobilize construction and support personnel.

X-10 Construction

On February 1, 1943, Du Pont started clearing the X-10 site, installing utility systems, and building the first temporary buildings, which consisted mostly of wooden barracks. In March, construction began on the six hot cells for plutonium and fission-product separation. The cells had thick concrete walls with removable slab tops for equipment replacement. The cell nearest the nuclear reactor housed a tank for dissolving uranium brought from the reactor through an underground canal; four other cells housed equipment for successive chemical treatments—precipitation, oxidation, reduction—of the uranium; a sixth cell stored con-

Remote controls for plutonium separations in 1944 at Oak Ridge.

taminated equipment removed from the other cells. A frame structure, abutting the cell walls, housed the remote operating gallery and offices.

Other structures rising at X-10 housed chemistry, physics, and health physics laboratories; machine and instrument shops; warehouses; and administration buildings. Because construction of the Y-12 and K-25 plants on the reservation also began in 1943, Du Pont had difficulty finding enough workers. It remedied the shortage by dispatching recruiters throughout the region.

That summer, about 3,000 construction workers completed about 150 buildings, at an initial cost of $12 million. Construction materials included 30,000 cubic yards of concrete, 4 million board feet of lumber, 4,500 gallons of paint, and 1,716 kegs of nails. Buildings went up rapidly, but needs so outran accommodations that a workers' cafeteria operated in a striped circus tent and an old schoolhouse served as office space and a dormitory.

Excavations for the foundation of the graphite reactor began in late April 1943; the reactor's thick concrete front face was in place by June; and the side and rear walls were finished in July. The National Carbon Company delivered graphite of the required purity to X-10, where Du Pont built a fabrication shop to machine graphite blocks to the desired dimensions. In September, a crew stacked the first of seventy-three layers of graphite blocks within the concrete shield to form a twenty-four-foot cube, and at month's end installed steel trusses to support the concrete lid capping the reactor. At the same time, the Aluminum Company of America began encasing sixty thousand uranium slugs in aluminum for the reactor, under government contract. Mounted in a building near the reactor, two of the world's largest fans sucked outside air through the reactor, then up a stack. The stack and the black building that housed the reactor (called the "black barn") were prominent features everyone was bound to notice when arriving at X-10 during the war.

In the meantime, Wigner had changed the cooling system design for the larger reactors to be built at Hanford, Washington, from helium to water. As a result, the air-cooled X-10 reactor was not truly a pilot plant for Hanford's water-cooled reactors. Instead, Du Pont officials viewed the hot cells of the separations building adjacent to the X-10 reactor as a pilot plant for similar facilities to be built at Hanford, and they considered development of chemical separations processes the most challenging mission at X-10.

Plutonium separation, in fact, challenged chemical engineers to design, fabricate, and test equipment for remotely transferring and evaporating liquids, dissolving and separating solids, and handling toxic gases. To meet this challenge, instrumentation was needed for remote measurements of volumes, densities, and temperatures in a hazardous environment. In addition, techniques to separate microscopic amounts of radioactive elements from volumes of liquid thousands of times larger had to be invented; the unknown effects of intense radiation on the solvents had to be identified and handled; and disposal of contaminated equipment and unprecedented volumes of radioactive wastes had to be addressed. These were a few of the challenges facing Clinton Laboratories personnel as work progressed at X-10 during the autumn of 1943.

The organization of Clinton Laboratories was in constant flux during the war. Scientists and technicians moved from Chicago to Oak

Personnel using a long rod
to push uranium slugs into
channels in the graphite
reactor at Oak Ridge.

Ridge to Hanford and Los Alamos as if they were in a revolving door.
Many members of the original research staff came largely from Chicago. The Du Pont Company brought its construction and operations
personnel to Oak Ridge for training, then moved them to Hanford.
Most Du Pont personnel came to X-10 from ordnance plants the company had constructed before 1943. Wartime employment at Clinton
Laboratories leveled off in 1944 at 1,513 scientists, technicians, and operating personnel, including 113 soldiers from the army's Special Engineering Detachment assigned to the Manhattan Project.

Organization of the laboratory proceeded in 1943, with Martin Whitaker as its director and Richard Doan as its associate director
for research. Reporting directly to Whitaker were research manager
Doan, Simeon Cantril (and later John Wirth) of the Health Division, and
Plant Manager S. W. Pratt, who brought many Du Pont personnel to
Oak Ridge. When its initial organization took shape, Clinton Laboratories had units for chemistry, separations development, health, production, engineering, and accounting, together with sections devoted to
physics and radiation biology.

Reactor Goes Critical

With a bit of unintended irony, Du Pont had completed the graphite reactor by Halloween in 1943, and Whitaker summoned Compton and Fermi from Chicago to witness its first operation. Three days later, workers began to insert thousands of uranium slugs into the reactor.

The sequence involved loading a ton or two of uranium, withdrawing control rods to measure the increase in neutron flux, reinserting the rods into the pile, loading another batch of uranium, then stopping again to assess activity levels, each time attempting to estimate when the reactor would achieve a self-sustaining chain reaction. A second shift continued this tedious procedure into the night, with Henry Newsom and George Weil plotting the flux curve. Weil had manipulated the control rod when Fermi brought Chicago Pile 1 to criticality the previous December, and he had come from Chicago to help achieve the same result in Oak Ridge.

The day shift loaded nearly ten tons of slugs, and the night shift set out to beat this record, working at both ends of the scaffold elevator at the reactor's face under the supervision of Kent Wyatt. In the middle of the night, Newsom and Weil, in the plotting room, recognized that one more batch of slugs would bring the reactor to the critical point, and they stopped the loading.

Before dawn on November 4, Louis Slotin drove to the guest house to awaken the two Nobel laureates, Compton and Fermi, known by the aliases Holley and Farmer in Oak Ridge. In the dark, they raced down Bethel Valley Road to witness the reactor going critical at five that morning. Scientists aware that the world's first powerful nuclear reactor had gone critical were thrilled. John Gillette, a Du Pont engineer on the graveyard shift that had loaded the last twenty tons of uranium slugs, was "too pooped to care."

Arthur Rupp of the Engineering Division had been dubious of Wigner's theoretical calculations of the amount of heat that uranium would emit during fission. To test the computations, he and his colleagues calibrated the airflow through the reactor and installed temperature, humidity, and barometric instruments. They then compared the uranium fission rate with the amount of heat released. When the experimental value proved nearly the same as the theoretical prediction, Rupp's skepticism ended. "I knew then," he said, "the atomic bomb was going to work!"

As Wigner and Alvin Weinberg at Chicago had predicted during the design phase, the reactor had gone critical when about half its 1,248 channels were loaded. Initially called the X-10 or Clinton Pile, it became known as the graphite reactor, so well designed that it worked with few operational difficulties throughout twenty years of service.

Near the end of November 1943, the graphite reactor discharged the first uranium slugs for chemical separation. By year's end, the chemists had successfully extracted 1.54 milligrams of plutonium from the slugs and dispatched them to Chicago, by secret courier, in a container resembling a penlight. Blocking empty channels in the graphite (to concentrate the cooling air) allowed an increase in the reactor's thermal power to 1,800 kilowatts in early 1944. Subsequent airflow modification, plus the installation of larger fans for cooling, permitted its operation at more than 4,000 kilowatts, nearly four times the original design capacity, with corresponding increases in plutonium production.

Plutonium Production

In February 1944, the first plutonium shipment went from Oak Ridge to Los Alamos. By spring, the chemists had improved the bismuth phosphate separation process to the point that 90 percent of the plutonium in the slugs was recovered. By early 1945, when plutonium separation ceased at X-10, the graphite reactor and separations plant had produced a total of 326.4 grams of plutonium, a substantial contribution to nuclear research and ultimately to weapons development.

In early 1945, Robert Oppenheimer urgently requested Clinton Laboratories to supply Los Alamos with large quantities of pure radioactive lanthanum, called RaLa, the decay product of radioactive barium-140. Clinton's chemists separated the first quantity of this isotope from the reactor's fission products in glass equipment in the chemistry laboratory. To obtain larger amounts safely, Martin Whitaker assigned Miles Leverett the job of designing, constructing, and operating a barium-140 production facility. With support from the Chemistry Division, Leverett, Charles Coryell, and Henri Levy met the schedule and Oppenheimer's requirements. "I believe," Leverett later speculated, "that this was the first production of a radioisotope on a large scale."

To assist with the design of Hanford's plutonium production reactors, many experiments were performed at the graphite reactor during 1944. One test involved laminated steel and Masonite radiation shields. The shield samples were set in an opening in the graphite reactor to allow study of the interactions between the samples and radiation. Brass, neoprene, Bakelite, rubber, and ordinary construction materials to be used at Hanford also were exposed to radiation in the graphite reactor for performance analysis. Because the Hanford reactors were to be water-cooled, tubes were installed in the graphite reactor to circulate water and observe its cooling and corrosive effects.

The conventional relationship between pilot plant and production plant existed between the Clinton Laboratories' hot cells and similar concrete structures built at Hanford. George Boyd, John Swartout, and other chemists from Chicago moved to Oak Ridge in October 1943, where they continued their investigations of plutonium separation processes and the properties of plutonium.

The Clinton experience indicated the bismuth phosphate carrier process was not entirely suitable for plutonium separation, but Seaborg's other process, using lanthanum fluoride, worked well. This process was incorporated into Hanford's separation facilities. So was the experience of hundreds of personnel trained at Clinton Laboratories.

Physicist John Wheeler worried that unwanted isotopes capable of stopping chain reactions would be found in the irradiated uranium. Like the boron and cadmium used in reactor control rods, the isotopes would have a large neutron-capture cross-section—that is, they might absorb enough neutrons to kill a nuclear chain reaction. This problem occurred at the first Hanford reactor during its trial run, a nasty surprise to Fermi and all concerned. After the chain reaction became self-sustaining, the reactor stalled. A few hours later, the reactor, for unexplained reasons, started again. Fermi and Wheeler suspected that the

Clinton Laboratories as it looked in 1946.

isotope xenon-135, which decays in roughly the same time that the reactor had shut down, was the culprit.

Urgent, around-the-clock efforts to measure the neutron-absorption cross-section of xenon-135 began at the Argonne and Clinton laboratories. Scientists worked forty hours at a stretch to separate xenon-135 from its parent iodine, place samples in the graphite reactor, and obtain rough estimates of its ability to capture neutrons, an ability measured in barns (from the folk idiom "big as the broad side of a barn").

They measured xenon-135 at four million barns; that is, tiny amounts of xenon could shut down large reactors, which would start again after the xenon decayed.

Clinton Laboratories was criticized for not detecting xenon's effects during earlier graphite reactor operations. A decline of reactivity resulting from xenon poisoning had occurred in the graphite reactor, but the reactor's conservative design had overcome the poisoning effects.

The reactor did not shut down, and the staff did not think that the decline in reactivity would pose a problem.

Fortunately, at Hanford the Du Pont engineers also had designed reactors larger than necessary. This overdesign allowed the insertion of sufficient uranium fuel to overcome xenon's poisoning effects and to continue production of the plutonium later used in the "Trinity" test in July 1945 and in the bomb that devastated Nagasaki on August 9.

Battle of the Laboratories

Announcing the bombing of Hiroshima on August 6, President Harry Truman cited the weapons facilities built at Oak Ridge, Hanford, and Los Alamos, commenting, "The battle of the laboratories held fateful risks for us as well as the battles of the air, land and sea, and we have now won the battle of the laboratories as we have won the other battles."

This news came as a surprise even to some employees at Clinton Laboratories. Before he heard the president's announcement, reactor operator Willie Schuiten did not believe coworkers who told him the reactor's work was tied to a new weapon. He later commented, "The people in charge really did a good job of keeping the project a secret."

Many Oak Ridge scientists, however, knew or surmised the purposes of the project. News of the bomb's success elated them, especially if they had relatives serving in the armed forces in the Pacific. One physicist commented that "we had helped to do a bold and difficult job, and had stopped a war in its tracks." He added, "That was enough for the moment. Second thoughts came later."

A few days later came Japan's surrender and the end of World War II. Staff members drifted about Clinton Laboratories, gathering and talking, seemingly bereft of energy. "Everyone," admitted one scientist, "felt a sense of disorientation, of slackness, of loss of direction."

The war's end had come while Clinton Laboratories was in the throes of a management overhaul. In July 1945, one month before the first atomic bomb was dropped, the University of Chicago—its war-related research complete—withdrew as the contract operator, and the army selected Monsanto Chemical Company as the new operator. This

major change, combined with the fact that many scientists planned to
return to the universities and their prewar research, raised a fundamental question: "What will become of the lab?"

Living with Peace

Winning the war left the staff of Clinton Laboratories with both pride
in their accomplishment and a sense of anxiety. Their prime task of
guiding the Hanford facility in producing and separating plutonium
for use in an atomic bomb had been accomplished on schedule. But
with this task successfully completed, the future looked uncertain. Could
Clinton Laboratories be as useful and productive in peace as it had been
in war? Would its scientists be content to remain in the hills of East
Tennessee, or would they opt to return to more cosmopolitan settings
in Chicago, New York, and California? Would the federal government
be willing to invest as much money in the peaceful uses of nuclear energy as it had in weapons production?

Although Clinton Laboratories had emerged from the shadow of
war as a heroic place, its future was uncertain. Impressed by its bucolic
atmosphere and its impressive record of accomplishment, Eugene Wigner
thought Clinton Laboratories did indeed have a future. In late 1944, he
drew up a plan for an expanded postwar laboratory for nuclear research with as many as 3,500 personnel and an associated school of reactor technology. Furthermore, he hoped he and his theoretical group
in Chicago would be transferred as a unit to Oak Ridge.

Even though that plan was never realized, Wigner persuaded some
of his staff in Chicago to move south, starting in May 1945 with Alvin
Weinberg. Wigner followed in 1946, marking the opening of a volatile
era in Oak Ridge. Like the rest of America and the world, Clinton Laboratories, whose energies and resources had been focused exclusively on
war, would have to learn to live with peace.

Chapter 2

High-Flux Years

High-flux conditions prevailed at Clinton Laboratories after the war, when surprising decisions affecting its future were made in St. Louis, Chicago, and Washington, D.C. At the federal level, management of the national laboratories shifted from General Leslie Groves and the Army Corps of Engineers to David Lilienthal and the newly created civilian Atomic Energy Commission (AEC). In Oak Ridge, the contract with Monsanto Chemical Company, the industrial operator for Clinton Laboratories, was not renewed. The University of Chicago, a possible replacement, failed to assemble a management team, resulting eventually in the selection of a new industrial contractor, Union Carbide Corporation. Clinton Laboratories became Clinton National Laboratory in 1947 and Oak Ridge National Laboratory in 1948. In short order, one unexpected event followed another.

Despite the management tumult, solid accomplishments in science and technology were achieved. Under the leadership of Eugene Wigner, Clinton Laboratories designed a high-flux materials testing reactor, the precursor of all modern light-water reactors, and experimented with the Daniels Pile, a forerunner of high-temperature gas-cooled reactors. The first of thousands of radioisotope shipments left the graphite reactor in 1946, initiating a program of immense value to medical, biological, and industrial science. New organizational units were formed to study biology, metallurgy, and health physics, and several solid scientific accomplishments were recorded in these fields before the departures of Wigner and Monsanto.

Management fluctuations proved a source of anxiety and despair among staff during the 1947 Christmas season. By the start of the new year in 1948, however, crucial management decisions assured the survival of Clinton—renamed Oak Ridge National Laboratory—as a national laboratory, with a much broader mandate for fundamental science than it had had during the war.

Monsanto's Management

During the war, security concerns required officials to refer to Clinton Laboratories by its code name, X-10. The personnel of Monsanto Chemical Company, the new operating contractor, continued this practice in the postwar years. The remote Appalachian location of Clinton Laboratories, along with its unpaved streets and Spartan living conditions, presented an easy target for ridicule. Metallurgical Laboratory personnel in Chicago called X-10 "Down Under," while Du Pont personnel labeled it the "Gopher Training School." In official telegrams, Monsanto's staff referred to Oak Ridge as "Dogpatch," taking their cue from a popular comic strip lampooning "hillbilly" Appalachian life. Such ill-concealed scorn did not augur well for either postwar Monsanto administration or Clinton research.

The choice of Monsanto as contract operator of Clinton Laboratories seemed logical because of the latter's focus on chemistry and chemical technology. Monsanto was also interested in becoming a key player in nuclear reactor development. Charles Thomas, Monsanto vice-president, was the driving force behind the company's entry into nucleonics. A famous chemist, Thomas had established a laboratory at Dayton, Ohio, that Monsanto purchased in 1936, making it the company's central research laboratory.

In 1943, General Groves gave Thomas and Monsanto responsibility for fabricating nuclear triggers at the Dayton laboratory. When Thomas also agreed to supervise the operation of Clinton Laboratories in 1945, he merged both facilities into a single project and appointed himself project director, although he kept his main office at Monsanto's corporate headquarters in St. Louis.

When Whitaker and Doan left Oak Ridge, Thomas decided to estab-

General Leslie Groves and David Lilienthal discuss the 1946 transfer of Oak Ridge from the army to the Atomic Energy Commission.

lish a dual directorship at Clinton Laboratories—one for administration and another for research and development. Both directors would report to him. For executive director in charge of general administration and operations, he selected James Lum, who had assisted him in managing the Dayton laboratory. As Lum's assistant, he brought in Prescott Sandidge, who had managed Monsanto phosphate and munitions plants.

Transferring sixty personnel to Oak Ridge from other Monsanto plants, Thomas reorganized Clinton Laboratories' administration. Among the new administrators were plant manager Robert Thumser, shop and instrument superintendent Hart Fisher, chief accountant Clarence Koenig, and superintendent of support services Harold Bishop. Because many scientists returned to universities at the end of the war, Thomas and the Clinton staff also had to recruit replacements. Among the new staff members were Walter Jordan, P. R. Bell, and Jack Buck, who came from the radar laboratory at the Massachusetts Institute of Technology, and Ellison Taylor, Henry Zeldes, Harold Secoy, and Frank Miles, who came from the closed wartime laboratory at Columbia University.

Frederick Seitz, Eugene Wigner, Alexander Hollaender, and James Lum, who managed Clinton Laboratories in 1946, hike a ridge overlooking the laboratory.

The staff at Clinton Laboratories reached 2,141 in 1947 under Monsanto's management, making facilities expansion imperative. A moratorium on new building construction during 1946 and 1947, while the Laboratories' future was debated in Washington, caused personnel and equipment to be moved into empty buildings at the Y-12 plant, which was shifting its focus from the electromagnetic separation of uranium-235 to precision machining of weapons components.

Expecting Clinton Laboratories to build the nation's first peacetime research reactor and its first electric power-generating reactor, Thomas courted Eugene Wigner, bringing him from Princeton to Oak Ridge several times during late 1945 to conduct seminars and consult on reactor designs. In early 1946, he lured Wigner into a year's leave from Princeton to become Clinton Laboratories' research and development director by promising to relieve him of administrative duties, which Thomas assigned earlier to James Lum. Wigner also acquired an assistant for the administration of research and development: Edgar Murphy, a scientist who had served as army major during the war in the Manhattan District office.

When his Princeton colleagues asked Wigner why he was going to "Dogpatch," he told them that, as one of the three major nuclear research laboratories in the United States, Clinton Laboratories would become important "in the life of the whole nation." As its research director, he intended to focus on science education by developing research reactors suitable for use at universities, establishing nuclear science training under his former graduate student Frederick Seitz, and coordinating scientific research with universities throughout the South.

"Only too much have both Chicago and Oak Ridge lived in the past on fundamental knowledge that has been acquired either before the war or at one of the other government research centers," Wigner observed. "As these wells begin to run dry, this situation becomes increasingly unhealthy." To avoid this pitfall, he thought the laboratories should nurture cooperative ties with universities, which would prove mutually beneficial.

Early in his tenure, Wigner outlined his weekly routine. On Mondays, he would remain in his office with an open door to hear staff advice and grievances. On "Holy" Tuesdays, he would vanish, pursuing his own research to "keep my knowledge alive." Although he avoided committee meetings to the extent possible, the remainder of the week he would attend to duties, circulating through Clinton Laboratories to discuss scientific and administrative problems. "We'll have long arguments just as you are having them now with each other," he warned the staff, "and I fully expect to be wrong in most of them—that is from Wednesday to Friday."

High-Flux Designs

When Wigner arrived as research director, Clinton Laboratories was embarking on the design of new reactor types. Intense investigations began of a high neutron flux reactor useful for testing materials, and of the gas-cooled Daniels Pile for demonstrating nuclear energy's value for electricity production. Chemists at Clinton Laboratories also initiated research aimed at a high-flux homogeneous reactor.

Wigner devoted most of his attention to the high-flux reactor, subsequently renamed the materials testing reactor. Its chief function was to provide intense neutron bombardment for testing materials to be used in future reactors. It was a reactor designer's reactor and provided the most intense neutron source at the time.

Initial designs called for use of enriched uranium fuel with heavy water in the interior lattice serving as the moderator and the exterior cooled by ordinary (light) water. Wigner and Alvin Weinberg, appointed by Wigner to be Lothar Nordheim's successor as chief of physics, at first concluded that heavy water could severely reduce the neutron flux.

Squeezing laboratory-made heavy water out of the design, they selected more readily available river water as both moderator and coolant. Instead of uranium rods canned in aluminum as in the graphite reactor, the fuel element or core would be uranium sandwiched between aluminum cladding or plates. To ensure a high thermal neutron flux for research, the plates were surrounded by a neutron reflector made of beryllium. In time, this design served as the prototype for many university research reactors and, in a sense, for all light-water reactors that later propelled naval craft and generated commercial power.

Miles Leverett and Marvin Mann headed a team of scientists and engineers undertaking the materials testing reactor design at Oak Ridge. About sixty personnel became involved in the design over nearly six years.

Wigner's best-known contribution was the curved design of the aluminum fuel plates in the reactor core. These plates were placed parallel to one another with narrow spaces set between them to create channels that would carry the cooling water; the reactor's power was set largely by how much water flowed past the fuel plates. Concern arose that intense heat might warp the plates, bringing them in contact and restricting coolant flow. After pondering this potential problem, Wigner directed that the plates be warped, or curved, in advance to improve their structural resistance to stress. Because the warped plates could bow only in one direction, they would not constrict water flow.

Adjacent to the materials testing reactor, Leverett's team planned to construct a plant to reprocess nuclear fuel, using the precipitation process developed during the war. Chemists John Swartout and Frank Steahly recommended that the "25 solvent-extraction process" replace the more expensive precipitation process. Their recommendation was accepted. Solvent extraction eventually became the standard method worldwide for reprocessing spent nuclear fuel.

Monsanto's principal concern was the Daniels Pile, named for Farrington Daniels, who, at the Chicago Metallurgical Laboratory in 1944, had designed a reactor with a bed of enriched uranium pebbles moderated by beryllium oxide and cooled by helium gas. Some called it the pebble-bed reactor. In May 1946, the Manhattan District directed Monsanto to proceed with the design, leading to the construction of an experimental Daniels Pile to demonstrate electric power generation.

To accomplish this task, Monsanto brought Daniels from the University of Wisconsin as a consultant. The company also recruited engineers from industry and brought them to Clinton Laboratories, where they formed a Power Pile Division headed by Rogers McCullough. This division identified materials suitable for high-temperature reactors and developed pressure vessels and pumps, piping, and seals for high-pressure coolants; it also studied heat exchanger designs.

Recruited largely from outside Clinton Laboratories, however, the Power Pile Division was never fully integrated into the organization. The project, moreover, encountered numerous design problems. Critics of the Daniels Pile contended that it would never become a practical power-generating reactor and that building a demonstration project wasted time and resources. After all, Logan Emlet and operators of the graphite reactor had demonstrated power production with a toy steam engine and generator that used heat from the air-cooled graphite reactor. Why, critics said, should researchers pursue a more complicated and expensive power-production strategy?

Such criticism caused high-level support for the Daniels Pile to wane by 1948. It was never constructed, and Daniels, as a professor at the University of Wisconsin, would later gain renown as a national expert on solar energy.

Atoms for Health

Distribution of the radioisotopes produced at the graphite reactor for biological and industrial research rapidly became the most publicized activity at Clinton Laboratories in the postwar years. Orders began arriving soon after a radioisotope catalog was published in the June 1946 issue of *Science* listing the radioisotopes that Waldo Cohn and his group could prepare and ship.

On August 2, 1946, Wigner stood in front of the graphite reactor to hand the first peacetime product of atomic energy, a small quantity of carbon-14, to an official of a cancer research hospital in St. Louis, home of Monsanto Chemical Company. Soon, nearly fifty different radioisotopes were regularly shipped from Clinton Laboratories. In 1947, to handle their production and distribution, Logan Emlet of Operations established an Isotopes Section headed by Arthur Rupp; as the program expanded, it later became an Isotopes Division, headed by Rupp, John Gillette, and others.

One of the earliest cases of technology transfer from Clinton Laboratories came as a spin-off of the radioisotopes program. Abbott Laboratories located its original radiopharmaceutical production plant in Oak Ridge near the source of the radioisotopes. The plant moved to Chicago in the 1960s when ORNL ceased commercial production of most radioisotopes.

Today, one out of three patients admitted into a hospital in the United States receives diagnostic or therapeutic radiation treatment. Radiation, in brief, is a fundamental tool of modern medicine, and it is inconceivable to envision patient care without it.

Much of the foundation of radiation medicine was established during the early years of isotope development that took place at Clinton Laboratories. Yet, part of that progress was attained through experimental radiation treatments conducted on human subjects without their consent, including a fifty-five-year-old black employee of the laboratory who in 1945 was injected with plutonium after suffering multiple injuries in a near-fatal automobile accident.

Medical ethics were less demanding then than now, and these experiments did enable scientists and doctors to understand better the consequences of radiation exposure, including its therapeutic value. Nevertheless, an unwillingness to obtain informed consent raises serious moral and social issues that only now—with the full disclosure policy unveiled by Secretary of Energy Hazel O'Leary—are entering center stage of the public policy debate.

High-Flux Organization

Like most new managers, Wigner sought to sharpen the organization's mission and improve its performance. He made both minor changes, such as the appointment of Edward Shapiro as chief of technical libraries, and major changes involving the formation and staffing of new divisions. Thinking solid-state physics a key to reactor design, Wigner established a small group for solid-state studies in the Physics Division under Sidney Siegel and Douglas Billington; he formed a new division to investigate the effects of radiation on metals; and he persuaded Monsanto executives to consolidate and augment staffing of the machine shops that supported the research projects.

During the war, small machine shops scattered among several divisions had provided the tooling, finishing, and precision machine work required for scientific experimentation. In 1946, Wigner urged that these shops be merged into groupings comprising at least two hundred craftsmen. After some resistance to the suggestion, Executive Director James Lum established the central research shops in 1947 and imported Paul Kofmehl, a Swiss craftsman, as superintendent with Earl Longendorfer as his assistant.

Skilled craftsmen, who machined the hardware for the reactors and other projects, worked in the research shops. They acquired apprentices in the ancient tradition of the crafts and supplied scientists and engineers with the unique equipment and tools they required. As the work load increased, the research shops evolved into central machine shops and eventually became the Fabrication Department in the Plant and Equipment Division under the supervision of Robert Farnham. The shops even included an old-fashioned Tennessee blacksmith, Miller Lamb, who fabricated lead bricks for radiation shielding and produced customized nuts, bolts, and metal parts. Nearly a quarter century after Lamb retired in 1969, ORNL personnel still passed his handiwork every day: He forged the ladder rungs on the smokestacks at the laboratory.

In 1945, Miles Leverett purchased a secondhand mill to roll, cast, and forge reactor fuel elements and metal parts. He also recruited metallurgists for materials research. Declaring that "an integrated program on the properties and possibilities of materials from the structural and nuclear point of view is greatly to be desired," in 1946 Wigner hired William Johnson from Westinghouse as a consultant on the formation of a Metallurgy Division. Johnson recruited a half-dozen metallurgists to form the division under the leadership of John Frye, Jr.

Metallurgists faced the challenge of fabricating reactor components of uranium and aluminum alloys, beryllium, zirconium, and other exotic metals, and conducted intensive research into the functioning of metallic elements under high temperatures and radiation stress in reactors. Starting with fewer than a dozen staff members, the Metallurgy Division increased in time to as many as three hundred people. In 1952, Frye organized a group under John Warde as a ceramics laboratory. It fabricated crucibles, insulators, and fuel elements; it also customized parts for reactors, purified graphite for molds, developed vitreous enamels, and conducted significant ceramics research. It employed scientists and engineers and also a practical potter or two to make molds. From these modest beginnings, ORNL would become a world center for metallic alloys and ceramics research.

High-Flux Biology

Just as the atom's nucleus captivated physical scientists, the living cell was the center of attention for life scientists. The graphite reactor furnished a variety of radioisotopes to supplement the few tracers used earlier to investigate life processes within cells. Radioactive tracers brought about a revolution in the life and medical sciences, leading to a new understanding of metabolic processes and genetic activities. These developments in biological sciences and the need to better understand the effects of radiation on human health and the environment led Wigner to expand the biology and health physics organizations.

When John Wirth, head of the Health Division, returned to the National Cancer Institute in September 1946, Wigner and Lum split the Health Division into two new research sections plus a medical department headed by physician Jean Felton and later by Thomas Lincoln. In October, Wigner recruited Alexander Hollaender to form and head a Biology Division. Hollaender had received degrees in physical chemistry from the University of Wisconsin. At the National Institutes of Health, he had studied the effects of radiation on cells and the use of ultraviolet light to control airborne diseases.

Hollaender's initial research plan at Clinton Laboratories called for the study of radiation's effects on living cells, including such cell constituents as proteins and nucleic acids. Beginning with a few radiobiologists studying microorganisms and fruit flies in crowded rooms behind the dispensary, Hollaender initiated a broad program that would make his division the largest biological laboratory in the world. Hollaender would successfully unite fundamental research in the biological sciences with physics, chemistry, and mathematics, and would recruit widely to staff the initial research units: biochemistry, cytogenetics, chemical physiology, physiology, and radiology. William Arnold, Waldo Cohn, Richard Kimball, Elliot Volkin, and William and Liane Russell were among the Biology Division's most respected staff members, a group that included seventy scientists and technicians by 1947. Lacking space at the X-10 site, the new division moved into vacated buildings at the Y-12 plant.

The Biology Division's genetic experiments conducted under the supervision of William and Liane Russell, who used mice to identify the long-term genetic implications of radiation exposure for humans, attracted the most public interest. Among the division's early scientific accomplishments, however, Hollaender took special pride in the discovery by William Arnold of the electronic nature of energy transfer reactions in photosynthesis, the discovery by Waldo Cohn and Elliot Volkin of the nucleotide linkage in RNA, and the discovery by Volkin and Larry Astrachan of messenger RNA. The Biology Division's greatest long-term influence on science, however, may have come from its cooperation with the University of Tennessee–Oak Ridge Graduate School of Biomedical Sciences and with universities and research centers throughout the nation and the world.

The second division split from the old Health Division in 1946 was Health Physics, with K. Z. Morgan as its director. Health Physics soon included seventy personnel engaged in service, research, and education largely within Clinton Laboratories. The service section provided area monitoring and furnished personnel qualified to use improved radiation detection devices. Early research included studies of radioisotopes discharged into river systems, estimates of thermal neutron tolerances, and development of new methods to detect radiation.

William Russell conducted pioneering studies at the laboratory of the effects of radiation and chemicals on mice.

In 1944, Oak Ridge health physicists trained personnel responsible for radiation protection at Hanford. They continued this schooling at Oak Ridge until 1950, when AEC established fellowships for graduate study at Vanderbilt and Rochester universities. The army, navy, and air force also sent personnel to receive health physics training at Oak Ridge. In addition to its land-based monitoring efforts, the Health Physics Division used boats to measure radioactivity entering the Clinch River from White Oak Creek and airplanes to measure radioactivity in the air above Oak Ridge. As a result, the division was said to have its own "army, air force, and navy."

High-Flux Education

In late 1945, Martin Whitaker met with University of Tennessee officials to discuss a science education partnership that would allow young scientists to complete graduate studies at the university while working at Clinton Laboratories. This program was the precursor of an extensive cooperative graduate program with the University of Tennessee that continued and expanded.

In 1946, the Oak Ridge Institute of Nuclear Studies, a nonprofit corporation of fourteen southeastern universities, was chartered, with William Pollard as its director. In 1947, the institute became a government-owned, contractor-operated facility of AEC. Under its aegis, the Clinton Laboratories Biology Division trained scientists in the use of radioisotopes. Later, the institute opened a clinical facility to use radioisotopes for cancer treatment.

In 1949, the institute received support from AEC to open the American Museum of Atomic Energy in a wartime cafeteria building. In 1974, the museum, renamed the American Museum of Science and Energy, moved into a new building adjacent to the corporate headquarters of the Oak Ridge Institute of Nuclear Studies, which itself had been renamed Oak Ridge Associated Universities and had nearly fifty sponsoring members.

Universities that joined the institute were invited to use Clinton Laboratories' scientific facilities. Under the management of Russell Poor, the institute began a program for faculty research at Clinton Laboratories in the summer of 1947 with two participants. That number increased to seventy by 1950, a level maintained for many years. Supplementing this research program were traveling lectures and seminars conducted by Clinton scientists at the participating universities. The resulting interaction between these scientists and university faculties, along with faculty and student use of research equipment available at Clinton Laboratories, contributed significantly to the spectacular growth in graduate science education throughout the Southeast during the postwar years.

High-Flux Training

In August 1946, Eugene Wigner opened the Clinton Training School with Frederick Seitz as its director. Although Wigner envisioned it as a small postdoctoral seminar in nuclear technology, more than fifty people from the military, industry, and academia enrolled. Among the first participants were Herbert MacPherson, Sidney Siegel, John Simpson, Everitt Blizard, Douglas Billington, and Donald Stevens, all of whom subsequently became renowned for their contributions to science. The most famous graduate, however, was Captain Hyman Rickover of the U.S. Navy.

William Pollard, *center*, and Gould Andrews of the Oak Ridge Institute of Nuclear Studies welcome Eleanor Roosevelt to Oak Ridge.

The navy had first provided Wigner and Szilard funding for nuclear experiments in 1939. During the war, navy scientists developed a thermal diffusion process for separating uranium isotopes; the S-50 plant in Oak Ridge was built during World War II for this purpose. Navy interest in using nuclear energy for ship propulsion continued, and in early 1946 Philip Abelson of the navy research team spent several months at Clinton Laboratories studying Wigner's approach to reactor design. In May 1946, Admiral Chester Nimitz assigned five navy officers and three civilians to Oak Ridge. The officers were Hyman Rickover, Louis Roddis, James Dunford, Raymond Dick, and Miles Libbey. Rickover later recalled his Oak Ridge experience:

> When I started at Oak Ridge in 1946, there were four other naval officers along with me and three civilians. Each was sent to Oak Ridge individually, and each started working on his own. . . . As soon as I got to Oak Ridge, I realized that if we ever were going to have atomic powerplants

in the Navy, I would have to assemble these people and train them as a group. And I used a very simple expedient; I arranged to write their fitness reports, so once they knew I was writing their fitness reports, they started paying attention to me. So once I did that, then I was able to weld them into a team and teach them specialized duties in order to get ready for building a submarine plant. Well, the first attempt at building a powerplant at Oak Ridge was a civilian one, and it failed, then unofficially I persuaded the people, the engineers, and the scientists, who were engaged in that enterprise, without any formal permission, to start working on a submarine plant, and they did this for a while. Meanwhile, I advised the Chief of the Bureau of Ships to retain this group of trained people together, and as soon as we came back to Washington, to have us start working on a submarine plant.

Under Rickover's exuberant direction, navy officers, who were enrolled in the training school, attended every seminar, interviewed every scientist willing to talk, and wrote numerous reports that became the paper foundation of the nuclear navy. Rickover later chose the pressurized-water reactor proposed by Alvin Weinberg to propel the nuclear ships built by the navy. Legends about Rickover's activities at Clinton Laboratories still abound. For example, he sometimes elicited information from scientists by introducing himself by saying: "Hello, I'm Captain Rickover; I'm stupid," a salutation that recipients accepted at face value only at great personal risk to their careers.

With the end of Monsanto management and the return of Wigner and Seitz to their universities in 1947, the Clinton Training School ceased to exist. Despite its brief tenure, the school was responsible for launching a long and fruitful relationship between the navy and ORNL. Rickover entered into several nuclear design contracts with the Laboratory, and he often employed Oak Ridge scientists, such as Theodore Rockwell, Frank Kerze, and Jack Kyger, on navy projects. Everitt Blizard, a civilian who had accompanied Rickover to Oak Ridge, remained at the laboratory, where he supervised investigations of reactor shielding. For years, Rickover provided Alvin Weinberg with unsolicited advice on the Laboratory's management, which Weinberg received with both grace and steeliness. He also strongly supported the formation and subsequent educational work of the Oak Ridge School of Reactor Technology housed at ORNL between 1950 and 1965.

Research and Regulations

By his own account, Wigner's most troublesome problems as research director emanated from the army bureaucracy. In the postwar years, the army continued its wartime security policies. Such meddlesome oversight made the exchange of scientific data with Hanford and Los Alamos difficult for Wigner and his research staff. This and similar problems caused Wigner to have several confrontations with army authorities, notably Colonel Walter Leber.

Colonel Walter Leber had replaced Captain James Grafton as the army representative for Clinton Laboratories in May 1946, and he hired a large number of people to monitor the latter's activities. His office staff included twenty-two people to inspect construction and administration, three to investigate security breaches, and twenty-nine to examine research and development.

This large group audited even minor details, down to the book titles ordered by the library. Its actions soon alienated both Clinton scientists and Monsanto executives. James Lum, for example, strenuously objected to Leber's efforts to "interfere and assume responsibilities which are reserved only for Monsanto under the present contract." To reduce confusion and improve communications, Lum and Wigner asked Edgar Murphy, formerly an army major, to serve as a liaison with Leber's staff.

Tensions continued, however, notably in the case of critical experiments Wigner wished to undertake to test the use of beryllium as a neutron trap or reflector. He encountered a catch-22 situation created by Leber's interpretation of a regulation the army had imposed after Louis Slotin lost his life during a critical experiment at Los Alamos. Wigner insisted the tests were completely safe, but Leber required that the debilitating regulations, which brought the tests to a virtual standstill, be meticulously observed. Only after review at the highest level were the experiments allowed to continue. Such delays discouraged Wigner and in time caused him to return to university life.

High-Flux Science

"Speaking as individuals who have been interested in radiation effects on solids since the conception of the first large reactors," Wigner and Frederick Seitz wrote, "we find it gratifying that a phenomenon which originated as a pure nuisance promises to provide us with useful information about the solid state in general and about many of the materials we use every day."

By "nuisance" they meant the swelling and distortion of graphite under the bombardment of neutrons from nuclear fission, an effect predicted by Wigner and thus called the "Wigner disease." Concern about the impact of this "disease" on the graphite reactor at Oak Ridge and similar reactors at Hanford stimulated intense interest in solid-state physics at Clinton Laboratories and elsewhere in the postwar years. This fascination played a role in Wigner's formation of the Metallurgy Division and in his personal attention to neutron-scattering experiments and zirconium investigations.

Although aluminum had served as cladding for uranium in the graphite and other early reactors, it was not suitable for use in the high-temperature reactors designed in the late 1940s. Metallurgists considered substituting zirconium, a metal that resists corrosion in water at high temperatures. Zirconium, however, seemed to have an affinity for absorbing neutrons, ultimately "poisoning," or slowing, nuclear reactions.

In 1947, Wigner authorized a group of Clinton researchers to study this problem. Wigner devised a "pile oscillator" to move materials regularly in and out of a reactor. Using a washing machine gearbox to power the oscillator, Herbert Pomerance later that year discovered zirconium's affinity for neutrons had been vastly overstated because of its contamination by the element hafnium, which had a much greater poisoning effect.

Zirconium minerals have traces of hafnium, which is nearly identical in chemical characteristics to zirconium, making economical separation of the two difficult. With funding from Captain Rickover and the navy, many laboratories investigated ways to separate the two elements. In 1949, chemical technologists at the Y-12 plant under the direction of Warren Grimes developed a successful separation technique and scaled it to production under the direction of Clarence Larson, then superintendent of Y-12.

Zirconium alloys became essential first to the navy's reactors and later to commercial power reactors. Zirconium rods filled with uranium pellets made up the fuel cores of nearly all light-water reactors, and hafnium was used in the control rods to regulate nuclear reactions.

As authorities on solid-state physics, Wigner and Seitz were intrigued by the interaction of radiation with materials, especially the neutron-scattering experiments of Ernest Wollan and Clifford Shull, which used a beam of neutrons from the graphite reactor. With a modified X-ray diffractometer installed at a beam hole of the graphite reactor, Wollan and Shull systematically studied the fundamentals of thermal neutron scattering. Experiencing difficulty in making sense of the diffuse scattering from various forms of carbon—diamond dust, graphite powder, and charcoal—they called on Wigner for advice. Shull later recalled:

> I well remember a discussion that Ernie and I had with Eugene Wigner, then the research director of the laboratory and a physicist of infinite wisdom and physical intuition, about this puzzling feature. After listening to our tale of woe and reflecting on the problem, he surprised us very much by calmly suggesting "maybe there is something new here, and maybe we have to relax our notions about conservation of particles." I can only say that I came away from that meeting with the feeling that Wigner had more faith in our experiments than perhaps Ernie and I had!

After a few months' additional experimentation, Wollan and Shull recognized that the consistency of their data had been distorted by spurious multiple scatterings in the specimens being investigated, an effect unfamiliar to them. This breakthrough allowed them to pursue their studies, which established neutron diffraction as a quantitative research tool fostering scientific knowledge of crystallography and magnetism. Their work built the foundation on which neutron-scattering programs developed throughout the world, including a neutron crystallography program under Henri Levy in the Chemistry Division at the laboratory. Although a half century has passed since the initial experiments, neutron scattering remains a fertile field of research.

Ernest Wollan and Clifford Shull performed some of the world's first neutron-scattering experiments at the graphite reactor in Oak Ridge.

High-Flux Management

In late 1945, the War Department drafted a bill to continue military control of atomic research and energy. Atomic scientists at Chicago and Oak Ridge vigorously opposed the measure and formed associations to lobby for civilian control. After a protracted political battle, the Atomic Energy Act of 1946 established civilian control under a five-member commission. With David Lilienthal, formerly chairman of the Tennessee Valley Authority, as its first chairman, AEC assumed control from the Manhattan District in January 1947. While this high-level political struggle was in progress, the disposition of the facilities built by the Manhattan District, including Clinton Laboratories, was at issue as well.

In early 1946, General Groves had appointed a committee of prominent scientists to plan the Manhattan District's nuclear activities and budget for 1947. This committee urged the expansion of research and development for both the production of fissionable materials and the

First AEC commissioners: *left to right*, William Waymack, Lewis Strauss, David Lilienthal, Robert Backer, and Sumner Pike.

advancement of nuclear power. On one hand, the committee suggested awarding contracts to university and private laboratories for unclassified fundamental research. On the other hand, the committee urged that national laboratories assume responsibility for classified research and for experiments requiring equipment too expensive or products too hazardous for a university to manage.

As the committee viewed it, each national laboratory should have a board of directors from universities in its region that would form associations to sponsor research and perhaps become the contracting operators. The committee initially recommended only two national laboratories, one at Argonne near Chicago and another serving the northeastern states. It expected the eventual formation of a national laboratory in California, but it ignored the Southeast and other regions.

Led by George Peagram and Isidor Rabi of Columbia University, universities in the Northeast campaigned to acquire a national laboratory. The Radiation Laboratory at the Massachusetts Institute of Technology had closed at the war's end, and the Substitute Alloy Materials Laboratory at Columbia University had been moved to the K-25 plant in Oak Ridge. Columbia and other northeastern universities urged the relocation of Clinton Laboratories to the Northeast, and some scientists at Clinton Laboratories liked the idea. More importantly, General Groves was amenable to it, and he selected an old army post on Long Island as the future site of Brookhaven National Laboratory.

In April 1946, the University of Chicago agreed to operate Argonne National Laboratory, with an association of midwestern universities offering to sponsor the research. Argonne thereby became the first *national* laboratory. It did not, however, remain at its original location in the Argonne forest. In 1947, it moved farther west from the Windy City to a new site on Illinois farmland. When Alvin Weinberg visited Argonne's director, Walter Zinn, in 1947, he asked Zinn what kind of reactor was to be built at the new site. When Zinn described a heavy-water reactor operating at one-tenth the power of the materials testing reactor under design at Oak Ridge, Weinberg joked it would be simpler if Zinn took the Oak Ridge design and operated the materials testing reactor at one-tenth capacity. The joke proved unintentionally prophetic.

Clinton Laboratories' rural ambiance did not please Robert Oppenheimer, Isidor Rabi, and James Conant, influential members of AEC's scientific advisory committee. Early in 1947, Oppenheimer declared that "Clinton will not live even if it is built up." Perturbed by this attitude, Charles Thomas of Monsanto demanded improvements in Monsanto's operating contract at Clinton Laboratories, in part as a sign that East Tennessee would be included in the federal government's postwar plans. On a no-profit, no-loss basis, the contract's chief attractions for Monsanto were the inside knowledge it provided of nuclear reactor advances and the public relations benefits it accrued for the company as a result of its patriotic efforts to protect the nation's security and advance the nation's technological capabilities.

Such virtues had their limits, especially when the war's outcome was no longer at stake. During the 1947 contract renewal negotiations, Thomas requested that Monsanto be allowed to increase its maximum fee for services from sixty-five thousand to one hundred thousand dollars a month. This request was not well received at AEC; moreover, Thomas's request to build the materials testing reactor near Monsanto's

Dayton laboratory or near its corporate headquarters in St. Louis rather than Oak Ridge was unacceptable. In May 1947, Thomas and Monsanto decided not to seek to renew the contract for operating Clinton Laboratories when it expired in June. The company, however, agreed to serve on a month-to-month basis until AEC secured another contract operator.

Loss of the contract at Clinton Laboratories did not mar Charles Thomas's career. In early 1948, he signed a contract to operate the new AEC plant at Miamisburg near Dayton, later named the Mound Laboratory. That same year, he was elected president of the American Chemical Society, and in 1951 he became president of Monsanto. His director at Clinton Laboratories, James Lum (apparently thinking he had become a "headaches" expert), left for Australia in August 1947 to build an aspirin factory. Thomas made Lum's assistant, Prescott Sandidge, Clinton Laboratories' executive director, pending final contract closure. Colonel Walter Leber, temporary director for the army, left in the summer of 1947 as well, later becoming Ohio River Division commander for the Corps of Engineers and governor of the Panama Canal Zone.

In the summer of 1947, when Wigner returned to "monastic" life at Princeton, Clinton Laboratories was left without a research director. Thomas decided to leave selection of Wigner's successor to the new contract operator. He requested that Edgar Murphy, Wigner's assistant, coordinate research pending selection of a new contractor and director.

Of his work at Clinton Laboratories in 1946 and 1947, Wigner later lamented: "Oak Ridge at that time was so terribly bureaucratized that I am sorry to say I could not stand it. The person who took over was Alvin Weinberg, and he slowly, slowly improved things. I would not have had the patience."

Black-and-Blue Christmas

Because the Argonne and Brookhaven laboratories would be operated by associations of universities, William Pollard and the Oak Ridge Institute of Nuclear Studies considered assuming Monsanto's contract. AEC, however, preferred that the University of Chicago resume its operation of Clinton Laboratories, and it announced in September 1947 that a contract would be negotiated with Chicago. The university thereby would become contract operator of both the Argonne National Laboratory and Clinton Laboratories, which was renamed Clinton National Laboratory in late 1947 while negotiations with Chicago were under way.

AEC was willing to enter a four-year contract with the university. Negotiations floundered, however, over the division of responsibilities between the university and AEC for personnel policies, salaries, auditing, and oversight. Moreover, the university decided to recruit a new director and management team for Clinton Laboratories, despite pleas for the return of Wigner. William Harrell, the university business manager, paraded prominent scientists to Clinton Laboratories for orientation; when offered the director's position, however, they all demurred. Near the end of 1947, Warren Johnson, wartime chief of the Chemistry Division, agreed to serve as the interim director, but only temporarily.

Concerned that AEC's research program might become too academic, Lilienthal established a committee of industrial advisors for AEC, and during a November visit to Oak Ridge, he discussed with Clark Center, manager of a subsidiary of Union Carbide Corporation at Oak Ridge, the possibility that the company assume management of Clinton Laboratories. Union Carbide managed the nearby Y-12 and K-25 plants, and it already had staff and offices in Oak Ridge that could easily add the Clinton Laboratories to their responsibilities. In addition, Union Carbide wanted to simplify its labor union relations. Workers at K-25 had joined a CIO union, while craftsmen at Clinton Laboratories had joined an AFL union. A December 1947 strike over wages and benefits at K-25, which were lower there than at Clinton Laboratories, threatened the company's tranquillity and productivity. By assuming management of Clinton Laboratories, Union Carbide hoped it could even out wages and benefits across Oak Ridge's industrial facilities and possibly abate labor tensions.

With Lilienthal ill and bedridden and other AEC commissioners on holiday excursions, Carroll Wilson, AEC's general manager, made the decision on Christmas Eve in 1947 to replace the University of Chicago with Union Carbide. At the same time, he decided to centralize all reactor development at Chicago's Argonne National Laboratory, transferring responsibility for the Oak Ridge high-flux materials testing reactor to Argonne. The day after Christmas, AEC commissioners concurred with these decisions. Wilson went to St. Louis to persuade Monsanto to hang

Eugene Wigner and Alvin
Weinberg came from Chicago to
manage research at Oak Ridge.

on at Oak Ridge an additional two months until Union Carbide could become sufficiently organized for the task. To James Fisk, director of research, fell the lot of carrying the message to Oak Ridge, where he received the welcome one would expect for a bearer of ill tidings.

Remembered in Oak Ridge long afterward as "Black Christmas," the shock came during the round of holiday parties. Reaction to the surprise was caustic. "Deck the Pile with Garlands Dreary," followed by several bawdy verses, reverberated through the hills. "It was rapid-fire and rough," admitted Lilienthal. He went on to say, "The people at Clinton Lab engaged in fundamental research felt they had been double-crossed, for we proposed to have Carbide & Carbon operate the lab (what was left of it, that is, minus the high-flux reactor), and this caused great anguish, not only among the chronic complainers but quite generally."

Laboratory staff declared the decisions represented a demotion from national laboratory status to a radioisotopes and chemical-processing factory. Leaders of the Oak Ridge Institute of Nuclear Studies fired messages to President Truman and AEC protesting the decisions as a blow to southern scientific aspirations.

This thinking ignored AEC's promise to continue fundamental research at the Clinton Laboratories, specifically in physics, chemistry, biology, health physics, and metallurgy. Rather than reducing its status, in January 1948, the official name became Oak Ridge National Laboratory, ending the use of Clinton, which was the nearest town to the facilities.

The first effect of these decisions on the laboratory was the transfer of the Power Pile Division, formed to study the Daniels pebble-bed reactor, to Argonne National Laboratory. Before leaving Oak Ridge, however, division researchers had begun studying Rickover's naval reactor. Harold Etherington, Samuel Untermeyer, and others in the group subsequently gained recognition with their designs of a reactor prototype for the atomic-powered *Nautilus* submarine and for an early breeder reactor.

AEC never released a precise definition of *national laboratory*. It granted the title, however, only to laboratories engaged in broad programs of fundamental scientific research, that had facilities open to scientists outside the laboratories and cooperated with regional universities in extensive science education efforts.

Oak Ridge clearly qualified for national laboratory rank, becoming one of three initial national laboratories. Argonne and Brookhaven laboratories, the other two national laboratories, were built in 1948 on new sites, making Oak Ridge the oldest national laboratory on its original site.

As these postwar maneuverings suggest, Oak Ridge, located in the Appalachian Mountains far from the bright lights of any metropolis, has had to prove from its earliest days that its location was appropriate for its purpose. Surviving in an environment of political and administrative intrigue had required institutional perseverance and ingenuity—qualities that would serve the laboratory's science and management well in the years ahead.

Chapter 3

Accelerating Projects

"Discovering how radiation does what it does to inorganic, organic, and living matter will benefit the entire world," declared biochemist Waldo Cohn as he speculated about ORNL's postwar research agenda. A vital question facing the laboratory in the years following World War II was how to obtain the means to pursue such research. After all, the laboratory's brief history had been devoted largely to supporting the development of the atomic bomb. Although scientists had touted peaceful applications of the atom, there were no assurances that the government would be willing to shift its administrative gears and resources to such research.

One answer to the laboratory's postwar research dilemma came from an unexpected source: investigations into nuclear-powered aircraft sponsored by AEC and partially funded by the U.S. Air Force. The plane never got off the ground, but the research directed toward this effort lifted to new heights the level of scientific knowledge in biology, genetics, and physics, and of technologies related to reactors, computers, and accelerators.

Flights of Fancy

Fantasies about the future applications of atomic power abounded just after World War II. Popular writing and art, which depicted atomic-powered ships, submarines, aircraft, trains, automobiles, and even farm tractors, stimulated public interest. These popular images came into sharp focus at Oak Ridge, where the laboratory participated in the development of nuclear-powered submarines, aircraft, and ships during the late 1940s and 1950s.

The application of atomic power to motion and travel became a centerpiece of the ORNL research program in the postwar era, and efforts to devise nuclear-powered transport, especially aircraft and submarines, involved many ORNL researchers. This research, in turn, contributed to the design of three nuclear reactors, the adoption of high-speed digital computers, and the acquisition of particle accelerators for nuclear physics. The efforts, moreover, fueled the laboratory's budget and staffing, both of which increased during the late 1940s and early 1950s under the management of its new contract operator, Union Carbide Corporation.

In February 1950, ORNL merged with the research divisions at Y-12—a move that strengthened and diversified the laboratory's research efforts. One direct result: projects designed to build reactor-driven machines that could travel over land, work underwater, and perhaps even fly. In the process, the laboratory hoped to turn the public's postwar atomic dreams into concrete demonstrations of atomic energy's potential contributions to society.

Spin-offs from research into atomic travel—in funding for biology, medicine, genetics, and computer science—ultimately would prove more useful than the primary research goal. Like a physically fit marathon runner who never reaches the finish line but finds value in trying, the laboratory—and the nation—could take pride in the important benefits derived from the effort.

Acquiring Y-12's research divisions increased ORNL staff by 50 percent, and by 1953, more than 3,600 people were working at the newly merged facilities. Moreover, the merger enabled the laboratory to acquire new divisions with strong applied science and heavy industrial technological capabilities. The laboratory also benefited from the transfer of state-of-the-art hardware, acquiring, for example, a von Neumann computer for data acceleration and several cyclotrons that could accelerate subatomic particles to unprecedented energies.

Accelerated Administration

Added responsibilities, personnel, and equipment created new challenges in ORNL management and administration. In late 1947, Union Carbide Corporation's Carbide and Carbon Chemical Company, later renamed the Union Carbide Nuclear Division, became ORNL's operations contractor. It enjoyed two advantages that would serve both the company and the laboratory well. First, the company's expertise in chemical engineering fit the tasks it would be asked to accomplish. Second, Union Carbide was no stranger to Oak Ridge. Since 1943, it had managed a large staff that operated the K-25 plant. In 1947, the government extended Union Carbide's responsibilities to the Y-12 production facilities. Thus, when AEC called on Union Carbide to oversee ORNL research activities in December 1947, it placed all Oak Ridge operations under unified management.

Union Carbide soon proved its mettle both to AEC and to ORNL personnel. Under the arrangement, Carbide executives—both at the corporation's international headquarters in New York City and at its regional headquarters in Oak Ridge—set laboratory work rules and pay scales. Virtually the entire laboratory staff went on Union Carbide's payroll. For its services, Union Carbide received a fixed fee from AEC that amounted to about 2 percent of the laboratory's annual budget.

Union Carbide appointed Nelson Rucker as the laboratory's new executive director. A graduate of Virginia Military Institute, Rucker joined Union Carbide in 1933 to manage a Carbide plant in West Virginia. He moved to Oak Ridge with Carbide in the early 1940s and remained there throughout the war. At the time of his appointment to the position of ORNL executive director, he was serving as Y-12's plant manager.

Rucker was responsible for overseeing the laboratory's daily activities. Playing a role comparable to that of a city manager, he saw that the institution functioned efficiently on a day-to-day basis, but he did not set its research agenda. That responsibility belonged to the laboratory's research director, a position that Union Carbide had much difficulty filling. Several prominent scientists rejected the position; Frederick Seitz, for instance, declined because the laboratory had lost its reactor projects to Argonne. In December 1948, Carbide asked Alvin Weinberg to take the job. He also declined, citing his youth and lack of experience, but agreed to become the associate director for research and development.

A biophysicist, Alvin Weinberg had studied the fission of living cells at the University of Chicago during the late 1930s. In 1941, he joined the Metallurgical Laboratory to investigate nuclear fission. As an assistant to Eugene Wigner, he participated in wartime reactor designs and, in May 1945, on Wigner's advice, moved to Oak Ridge to join the Physics Division, where he succeeded Lothar Nordheim as division chief in 1947.

Weinberg, whose ability to communicate his thoughts in writing was exceeded only by his rare scientific talent, captured the spirit of excitement and confusion that existed at Oak Ridge during the late 1940s when he wrote Wigner about his responsibilities as head of the Physics Division. "I feel in my new job a little bit like a trick horse-back rider at a circus," Weinberg told Wigner. "The idea seems to be to ride standing on three or four spirited horses, all of which are interested in going in different directions."

Limited work space constituted a major challenge facing Rucker, Weinberg, and other Union Carbide managers in the late 1940s. During the postwar turmoil, AEC suspended new construction and often deferred maintenance on existing structures, pending the government's decision on the laboratory's future. This wait-and-see attitude, which made sense given the uncertainties in Washington, D.C., continued while wartime frame structures swiftly deteriorated. The only new facilities erected at Clinton Laboratories between 1946 and 1948 were surplus army Quonset huts to relieve overcrowding, plus an electric substation and steam power plant built in futile anticipation that the proposed materials testing reactor would be built in Oak Ridge.

Overcrowding became serious in 1948 as the laboratory added new divisions, hired more personnel, and installed new equipment. These events led physicist Gale Young to complain:

> In accumulating technical people which it cannot use for lack of accommodations, I believe that the Laboratory has embarked on a course which is suicidal to itself and detrimental to the national interest. Until considerably more buildings have been erected, staff reductions, rather than increases, are in order.

In 1949, with the laboratory's future on a firmer, more stable footing, AEC budgeted $20 million for new construction, and Union Carbide initiated its "Program H" to replace wooden wartime structures with more permanent brick and mortar. In addition to paved streets, landscaped grounds, and renovated older structures, about 250,000 square feet of new office and laboratory space opened in the early 1950s.

Among the new facilities, three were of particular importance: Building 4500, ORNL's principal research building and administrative headquarters; a radioisotope complex consisting of ten buildings designed to process, package, and ship the laboratory's most valuable material exports; and a pilot plant for use in chemical processing. With this new construction, AEC and Union Carbide gradually hoisted the laboratory out of the East Tennessee mud.

Accelerated Development

AEC's 1947 decision to centralize reactor development at Argonne National Laboratory proved ill considered. Argonne's mandate from AEC to support navy reactor development and new programs for civilian power and breeder reactors strained its resources and capabilities. It therefore supported Oak Ridge's efforts to continue design and fabrication work in East Tennessee in order to concentrate on its own full plate of responsibilities in Chicago.

Taking advantage of this unexpected turn of events, in 1948, ORNL urged AEC to build the materials testing reactor on the Cumberland Plateau twenty miles from Oak Ridge. AEC, however, acquired a site in Idaho and, four years later, the newly built materials testing reactor at the Idaho National Engineering Laboratory began successful operation under the supervision of Richard Doan, formerly the research director at Oak Ridge. Two years before the reactor in Idaho began operation, however, ORNL had the world's first solid-fuel and light-water reactor at work in Oak Ridge. Despite the government's intentions to end reactor work at ORNL, the facility's deeply rooted efforts in the development of this technology refused to wither.

While designing the materials testing reactor in 1948, the laboratory built a small mock-up of the reactor to test the design of its controls and hydraulic systems. In 1949, Weinberg proposed installing uranium fuel plates inside the mock-up to test the reactor design under critical conditions. The AEC staff feared that Weinberg's initiative might become an opening wedge for a revived reactor program at Oak Ridge. "We have no plans," Weinberg reassured them, "to convert the critical experiment into a reactor." The mock-up experiment at Oak Ridge in February 1950 produced the first blue Cerenkov glow of a nuclear reaction underwater ever seen, and it provided superb training for those who were to serve subsequently as operators for the full-scale reactor in Idaho.

As its reactor program burgeoned, AEC relaxed its previous plans to centralize reactor development and construction at Argonne National Laboratory and the Idaho National Engineering Laboratory. In fact, AEC allowed ORNL to upgrade the mock-up's shielding and cooling systems. These improvements raised the system's capacity to three thousand thermal kilowatts, only one-tenth of the materials testing reactor's maximum power but still useful for experiments. Labeled the "poor man's pile" by Wigner, the mock-up formally became the low-intensity test reactor.

Experiments conducted there established the feasibility of the boiling-water reactor, which later became one of the design prototypes for commercial nuclear power plants. Operated remotely from the graphite reactor control room, the "poor man's pile" served ORNL until 1968 when AEC shut it down after a long, useful life.

Flying Reactors

With funds drawn largely from the U.S. Air Force, ORNL's major entrance into reactor development during the 1950s came through efforts to design a nuclear airplane. British and German development of jet engines at the end of World War II had given quick, defensive fighters an advantage over slower long-range offensive bombers. To address the imbalance, General Curtis LeMay and Colonel Donald Keirn of the air force urged development of nuclear-powered bombers. In 1946, they persuaded General Groves to approve air force use of the vacated S-50 plant near the K-25 plant in Oak Ridge to investigate whether nuclear energy could propel aircraft.

The initial concept called for a nuclear-propelled bomber that could fly at least 12,000 miles at 450 miles per hour without refueling. Such range and speed would enable conventional or nuclear weapons to be delivered via airborne bombers anywhere in the world. The aircraft,

Inside the core of the low-intensity test reactor.

however, would require a compact reactor small enough to fit inside a bomber yet powerful enough to lift the airplane into the air, complete with lightweight shielding to protect the crew from radiation.

Under air force contract, the Fairchild Engine and Airplane Corporation then established a task force at S-50 to examine the feasibility of nuclear aircraft and arranged with Wigner to receive scientific support from ORNL. Initial studies conducted by the Fairchild Corporation at S-50 showed promise, and in 1948, AEC asked the Massachusetts Institute of Technology to evaluate the feasibility of nuclear-powered flight. MIT sent scientists to Lexington, Massachusetts, for a summer's appraisal, and they reported that such flight could be achieved within fifteen years if sufficient resources were applied to the effort. In September 1949, AEC approved ORNL participation in an aircraft nuclear propulsion project; Weinberg was made project director and Cecil Ellis coordinator. Raymond Briant, Sylvan Cromer, and Walter Jordan later served as directors of the laboratory's Aircraft Nuclear Propulsion (ANP) project.

Soon after the laboratory acquired its nuclear propulsion project,

General Electric took over the work of Fairchild and relocated it from Oak Ridge to its plant in Ohio. Although some Fairchild personnel transferred to Ohio, about 180 remained in Oak Ridge to join the laboratory's aircraft project in May 1951. Among those who decided to stay in East Tennessee were François Kertesz, a multilingual scientist; Edward Bettis, a computer wizard before the age of computers; William Ergen, a reactor physicist; Fred Maienschein, later the director of the Engineering Physics and Mathematics Division; and Don Cowen, who headed the laboratory's Information and Reports Division.

Much of the laboratory's initial aircraft work focused on development of lightweight shielding to protect airplane crews and aircraft rubber, plastic, and petroleum components from radiation. Knowing a nuclear aircraft would never become airborne carrying the thick walls typical of reactor shields, Everitt Blizard and his team worked two shifts daily, testing potential lightweight shielding materials in the lid tank atop the graphite reactor. As the research progressed, however, the graphite reactor proved inadequate to meet the heightened level of research activity. To continue its shielding investigations, ORNL added two unique nuclear reactors to its fleet.

First, in December 1950, the laboratory completed its two-megawatt bulk shielding reactor at a cost of only $250,000. To build this reactor, the laboratory modified its earlier materials testing reactor design to create what became popularly known as the "swimming pool reactor." This reactor's enriched uranium core was submerged in water for both core cooling and neutron moderation. From an overhead crane, the reactor could be moved about a concrete tank, the size of a swimming pool, to test bulk shielding in various configurations. The laboratory later placed a ten-kilowatt nuclear assembly (named the *pool critical assembly*) in a corner of the pool to permit small-scale experiments without tying up the larger reactor.

The laboratory standardized this inexpensive, safe, and stable design, which became a prototype for many research reactors built at universities and private laboratories around the world. Upgraded with a forced cooling system in 1963, the bulk shielding reactor supplanted the graphite reactor (retired that year) as the laboratory's prime research reactor and proved extremely useful for irradiating materials at low temperatures.

A second laboratory reactor resulting from the nuclear aircraft project was the tower shielding facility completed in 1953. Cables drawn from steel towers could hoist a one-megawatt reactor in a spherical container nearly two hundred feet into the air. Because no shielding surrounded the reactor when suspended, it operated under television surveillance from an underground control room. Containing uranium and aluminum fuel plates moderated and cooled by water, the "tower" reactor helped scientists answer questions about the effects of radiation from a reactor flying overhead; it also helped researchers better understand the type and amount of shielding that would be needed aboard a nuclear aircraft.

Experiments indicated that a divided shield, consisting of one section around the aircraft's reactor and another around its crew, would comprise a combined weight less than a single thick shield blanketing the aircraft's reactor. Researchers, however, could never devise a reactor and shielding light enough to ensure safe flight. They even considered a "tug-tow" arrangement in which the crew and controls would be placed in a towed glider, separated from, yet tied to, the reactor by a long umbilical cable. The tower shielding facility reactor later was upgraded, and shielding experiments took place there until 1992, long after visions of a nuclear aircraft faded from memory.

Fireball Reactor

The bulk and tower shielding reactors were designed to test materials that might be used on a nuclear-powered aircraft. For the U.S. Air Force, improved materials represented a means toward an end: a nuclear-powered engine that could drive long-range bombers to takeoff speeds and propel them around the world. To achieve this goal, ORNL designed an experimental hundred-kilowatt aircraft reactor as a demonstration. This small reactor, operating at high temperatures, used molten uranium salts as its fuel, which flowed in serpentine tubes through an eighteen-inch reactor core. A heat exchanger dissipated the reactor's heat into the atmosphere. In 1953, the laboratory constructed a building to house this experimental reactor.

To contain molten salts at high temperatures within a reactor, the laboratory used a nickel-molybdenum alloy, INOR-8, fabricated at the

International Nickel Company. Able to resist corrosion at high temperatures while it retained acceptable welding properties, the alloy was commercialized as Hastelloy-N by private industry (an early example of technology transfer) to supply tubing, sheet, and bar stock for industrial applications. The aircraft reactor also compelled laboratory personnel to learn how to perform welding with remote manipulators and how to remotely disassemble molten-salt pumps. Laboratory researchers devised two salt reprocessing schemes, as well, to recover uranium and lithium-7 from spent reactor fuel.

The first test run of the aircraft reactor experiment took place in October 1954. The reactor ran at one megawatt for one hundred hours. Donald Trauger and other observers of the reactor's operations recall that the reactor core, pumps, valves, and components literally became "red hot." Completing the design, fabrication, and operation of such an exotic nuclear reactor in five years was considered a noteworthy event, and dignitaries such as General James Doolittle, Admiral Lewis Strauss, and Captain Hyman Rickover visited Oak Ridge to see the red-hot reactor in action.

Its success led the laboratory to propose additional study of this reactor concept and the design of a larger sixty-megawatt, spherical prototype, known as the "fireball reactor," to conduct more sophisticated experiments. Laboratory researchers, for example, asked what would happen if a plane turned upside down while irradiated molten liquid pulsated through the engine. More significantly, they wondered what would happen if a plane failed in midair or during takeoff or landing.

Three unique reactors were not the only hardware the laboratory acquired as a result of its nuclear aircraft project. The project helped justify the construction of a critical-experiments facility to test reactor fuels and a physics laboratory to study the effects of radiation on solid materials. It also advanced laboratory efforts to acquire its first nuclear particle accelerators and digital computers.

Because the success of nuclear flight depended on expensive and complex hardware on the ground, the laboratory benefited from being on the receiving end of a well-funded government project. However, its ability to take advantage of this situation also depended on the skill of its research and support staff and the managerial expertise of its leaders. Internal administrative adjustments, including the merger of the Y-12 research division with ORNL, also helped.

Y-12 Laboratories Acquired

By 1950, all parties—the government, ORNL, and the company—largely viewed Carbide's management of the laboratory as a success. Recognizing that staff loyalties resided with the laboratory, Carbide did not attempt to convert them to "company personnel." It eagerly identified and rewarded ambitious laboratory staff (elevating some to managerial positions), undertook sorely needed facility reconstruction and expansion, and fostered basic and applied sciences. "Carbide management has demonstrated," asserted one manager, "that first-rate basic research can be done in an industrial framework."

When Nelson Rucker, Carbide's executive director of ORNL operations, transferred to a carbide plant in West Virginia in 1950, a major reorganization ensued. Alvin Weinberg, formerly associate director, became the laboratory's research director, and Clarence Larson, formerly the Y-12 plant manager, became ORNL's new director.

A chemist from Minnesota, Larson had worked at the University of California's radiation laboratory before moving to Oak Ridge to become the Y-12 research director in 1943 and superintendent in 1948. An able manager and accomplished scientist, Larson strengthened and broadened ORNL's research activities.

Before Larson's appointment, Union Carbide considered moving ORNL to Y-12, where the Biology Division already occupied a building. By 1950, however, the chilling tensions of the Cold War and the heated battles of the Korean War sparked rapid expansion of nuclear weapons production, which increased the work load at Y-12 and K-25 and led to the construction of new gaseous diffusion plants at Paducah, Kentucky, and Portsmouth, Ohio.

As a result, space became precious at Y-12, and plans to move the laboratory there were aborted. Thus ORNL's acquisition of Y-12's three research divisions—Isotope Research and Production, Electromagnetic

Research, and Chemical Research—left everyone and everything in the same place. However, the administrative realignment meant that Y-12 researchers in these divisions began reporting to ORNL management.

Isotopes

By 1950, ORNL was distributing more than fifty different radioisotopes to qualified research centers. Cobalt-60, used for cancer research and therapy, was a prime isotope on the laboratory's distribution list. When ORNL began to ship isotopes overseas, AEC approved a cooperative arrangement between the laboratory and the Oak Ridge Institute of Nuclear Studies to train foreign scientists in radioisotope research. The laboratory's isotope research efforts were further advanced through the merger of Y-12's Isotope Research and Production Division with ORNL's Isotopes Division. This union added stable, nonradioactive isotopes to the laboratory's catalog.

The Y-12 stable isotopes program had emerged at the end of the war when the Y-12 staff ceased separating uranium isotopes for atomic weapons. Eugene Wigner then urged the continued use of some calutrons to separate the stable isotopes of all elements. "We should have as the very basis of future work in nuclear physics and chemistry, knowledge of the various cross-sections of pure stable isotopes," he urged. AEC approved Wigner's proposal, and a group led by Clarence Larson, Christopher Keim, and Leon Love began to separate various isotopes of stable elements.

Researchers at first used four calutrons salvaged from electromagnetic equipment. Stable isotope research and development required modifications to the calutrons, better understanding of the obscure chemistry of less common elements, spectroscopic analysis of nuclear properties, and advances in the use of isotopes as tracers.

Christopher Keim, a group leader, later recalled that copper isotopes were the first to be collected. "All that had to be done," Keim modestly explained, "was to put copper chloride into the charge bottle, heat it with uranium tetrachloride, lower the magnetic field, and space the collector slots to receive the copper-63 and copper-65 ion beams." Using enriched copper-65 as the source material for neutron irradiation, George Boyd, John Swartout, and colleagues positively identified nickel-65 as a nickel isotope with a half-life of 2.6 hours. This discovery represented the first use of calutron-separated stable isotopes in research.

Stable isotopes of iron, platinum, lithium, and mercury, for example, were separated and shipped to university, government, and industrial laboratories worldwide to aid basic research in physics, chemistry, earth sciences, biology, and biomedicine. They became especially valuable to medical science, for which they were converted into radionuclides used as tracers to diagnose cancer, heart disorders, and other diseases affecting human internal organs and bones. Contributing to basic scientific knowledge and enhancing the quality of human life, the laboratory's stable isotopes program continued to expand through the 1970s. At its height, the program generated more than $1 million annually in sales revenue.

Particle Accelerators

In 1950, the Y-12 Electromagnetic Research Division under Robert Livingston became ORNL's Electronuclear Division, switching from studies of calutrons to fundamental research on the formation and motion of ions in electric fields. The Electronuclear Division was also in charge of the cyclotrons used for particle acceleration. At the same time, Arthur Snell and colleagues in the Physics Division entered the particle acceleration field as well, using electrostatic accelerators popularly called "atom smashers."

Thus the laboratory, during the early 1950s, pursued two independent lines of particle acceleration—cyclotrons in the Electronuclear Division and electrostatic machines in the Physics Division. This hot pursuit of fast-moving subatomic particles was propelled by rapid postwar advances in the basic science of nuclear physics.

During the postwar years, exploration of the unknown particles and forces of atomic nuclei led to the discovery of subatomic particles smaller than neutrons, electrons, and protons. The study of these elementary particles emerged from nuclear science as a subfield labeled high-energy physics. Oak Ridge, as a national laboratory dedicated to fundamental research, was anxious to participate in the exploration of these microscopic particles.

Its research efforts had an abortive start in 1946, however, when the laboratory proposed to purchase a large betatron accelerator to join the hunt for elusive subatomic particles. This purchase required the approval of the army, and the resulting bureaucratic delays made the 160-ton betatron obsolete when it finally arrived. Saddled with an outdated piece of equipment, the laboratory sold it as surplus to another government agency. By 1948, however, the laboratory's nuclear aircraft program, with support from the U.S. Air Force, was inching down the runway. This project added impetus for accelerator research because of the need to understand the effects of radiation on shields and other materials that would be part of the aircraft.

In 1948, Arthur Snell, director of the Physics Division, asked Wilfred Good and Charles Moak to start an accelerator program using materials readily and inexpensively available at ORNL and Y-12. "The objective was clear," recalled Good. "Neutrons were the key to the new frontier of applied nuclear energy; to fully exploit neutrons, their behavior had to be thoroughly understood; and the Van de Graaff was the only known source of neutrons of precisely determined energies."

The Chemistry Division had acquired a two-and-a-half-megavolt Van de Graaff electron accelerator from the navy. Richard Lamphere of the Instrumentation and Controls Division converted it into a three-megavolt proton accelerator that could bombard lithium targets with protons to produce a stream of neutrons. This little Van de Graaff accelerator supported research for thirty years, its most important service to science coming when John Gibbons, Richard Macklin, and colleagues used it to confirm a theory that atomic elements originated through nucleosynthesis in the centers of stars.

To test radiation effects at energies lower than those generated by the Van de Graaff, the laboratory also acquired a Cockcroft-Walton accelerator, an early particle accelerator named for its inventors. The laboratory installed these first accelerators in an abandoned powerhouse.

In March 1949, Alvin Weinberg and Herman Roth of AEC met air force commanders and contractors to discuss priorities in the nuclear aircraft research program. After concluding that a five-megavolt Van de Graaff accelerator was needed, the air force agreed to purchase it if ORNL would construct a building to house it. First installed at the Y-12 plant, the five-megavolt Van de Graaff accelerator produced its first

One of ORNL's "atom smashers," the Van de Graaff accelerator fires charged particles into a
target

beam in 1951, making it the world's highest-energy machine of its kind. In
1952, ORNL completed a high-voltage laboratory building and moved
the three linear particle accelerators into it. A decade later, it added a
fifteen-megavolt tandem Van de Graaff accelerator to extend the energy capability of the existing machines and to accelerate ions heavier
than helium. More than thirty years later, ORNL physicists still view
this accelerator as a valuable research tool.

Cyclotron Acceleration

While Arthur Snell and members of the laboratory's Physics Division
concentrated on particle acceleration through direct-current high-voltage machines, Robert Livingston and the Y-12 electromagnetic team
pursued an independent course of achieving acceleration with cyclotrons. Invented in 1930 by Ernest Lawrence at Berkeley, cyclotrons had
two D-shaped electrodes (dees) in a large and nearly uniform magnetic
field. The dees operated at high electric potential and were alternately

positive or negative. They accelerated the charged particles (ions), and the magnetic field confined them to a circular orbit.

Cyclotrons were the forerunners of the giant synchrotrons of the 1990s, and during their sixty years of development they increased the energy of protons (nuclei of hydrogen atoms) from 1 million to 20 trillion electron volts. The cost of the machines also multiplied from $100,000 each to $10 billion each.

Having built calutrons during the war for the electromagnetic separation of uranium isotopes, Livingston and his associates at the Y-12 plant had abundant experience in calutron design and construction and took advantage of the unused electromagnets left after the war to advance their expertise. During the late 1940s and early 1950s, they built three cyclotrons to study the properties of compound nuclei and heavy particle reactions. The cyclotrons were identified by their diameters measured across the dees as the twenty-two-inch, sixty-three-inch, and eighty-six-inch machines.

Livingston's team built the twenty-two-inch cyclotron in the late 1940s to test how electromagnets in calutrons could be used and how high-current calutron ion source techniques could be applied to cyclotron functioning. The cyclotron served its purpose and later was doubled to forty-four inches for testing new ion sources, new beam-focusing methods, and new ways to increase the intensities of accelerated beams.

The eighty-six-inch cyclotron began operation in November 1950, performing radiation damage studies for the nuclear aircraft project. As the world's largest fixed-frequency proton cyclotron, it produced a proton beam four times more intense than that of any other cyclotron; its blue beam projected through the air as much as sixteen feet, visibly impressing visitors. Bernard Cohen, chief physicist for this machine, used it to study proton-induced nuclear reactions and to supply the isotope polonium-208 until a commercial source became available.

This was the era of hydrogen bomb development, and the question arose whether a powerful hydrogen bomb might ignite nitrogen in the atmosphere, causing a worldwide holocaust. To find the answer, AEC asked ORNL to build a cyclotron that would accelerate nitrogen ions. The laboratory asked Alex Zucker, who had recently received his doctorate from Yale University, to develop a source of multiply charged

Workers in 1950 completing construction of ORNL's eighty-six-inch cyclotron.

nitrogen ions. After successfully completing this task, he was directed to build a cyclotron to measure the cross-section of the nitrogen-nitrogen reaction and thereby determine whether the atmosphere would burn.

Built in eighteen months at a cost of three hundred thousand dollars, the cyclotron became operational in 1952. Zucker and his collaborators, Harry Reynolds and Dan Scott, soon demonstrated that a hydrogen bomb—despite its horrific consequences—would not ignite a chain reaction in the atmosphere. They then turned the cyclotron into a basic research instrument, the world's first source of energetic heavy ions, making the interactions of complex nuclei a new field of scientific investigation.

The laboratory's first cyclotrons were the most economical ones ever built because the Electronuclear Division used surplus electromagnetic equipment that required little modification. Because the Y-12 calutron tracks had been placed side by side in vertical formation, ORNL's cyclotrons were marked by their unique vertical mounting, in contrast to the horizontal position of the dees found at other laboratories. These pioneering cyclotrons helped advance the technology of high-beam currents, and they have since been the force behind the laboratory's versatile isochronous (variable energy) cyclotron completed in 1962, and still later the Holifield heavy-ion research facility completed in 1980.

Information Acceleration

The aircraft nuclear propulsion project, together with the reactors and particle accelerators developed to support it, generated immense quantities of scientific data that required rapid analysis. This need stimulated ORNL interest in electronic computers, which became available during the 1940s.

In 1947, Weinberg created a Mathematics and Computing Section within the Physics Division under the direction of Alston Householder, a mathematical biophysicist from the University of Chicago, who, in 1948, converted the section into an independent Mathematics Panel to manage the laboratory's acquisition of computers.

Before 1948, complex, multifaceted computations at the Y-12 and K-25 plants were done on electric calculators and card programming machines. Because of its participation in the nuclear aircraft project, ORNL obtained through the Fairchild project a matrix multiplier to solve linear equations. At the laboratory's urging, AEC also leased Harvard University's early Mark I computer. Householder and Weinberg insisted that the laboratory should acquire its own "automatic sequencing" computer to be used by staff scientists doing difficult computations for the nuclear aircraft project. The computer, they contended, also could serve to educate university faculty and researchers visiting the laboratory. When purchased, it became the first electronic digital computer in the South.

Householder and the laboratory's leadership were familiar with the pioneering work of Wigner's friend, John von Neumann, who had pursued experimental computer development near the end of the war for the navy. Admiral Lewis Strauss thought the navy needed computers to aid in weather forecasting, vital to ships at sea. At his urging, the navy in 1946 sponsored the fabrication of the first von Neumann digi-

ORACLE, or Oak Ridge Automatic Computer and Logical Engine, was the laboratory's
first large computer.

tal computer at Princeton University. After considering Raytheon and
other commercial computers, ORNL and Argonne National Laboratory
decided to build their own von Neumann–type computers, tailored
specifically to solve nuclear physics problems. ORNL engineers assisted
Argonne during the early 1950s in the design and fabrication of the Oak
Ridge Automatic Computer and Logical Engine. Its name was selected
with reference to a lyrical acronym from Greek mythology—ORACLE,
defined as "a shrine in which a deity reveals hidden knowledge."

Assembled before the development of transistors and microchips,
ORACLE was a large scientific digital computer that used vacuum tubes.
It had an original storage capacity of 1,024 words of 40 bits each (later
doubled to 2,048 words); the computer also contained a magnetic-tape
auxiliary memory and an on-line cathode-tube plotter, a recorder, and
a typewriter. Operational in 1954, for a time ORACLE had the fastest
speed and largest data storage of any computer in the world. Problems

that would have required two mathematicians with electric calculators three years to solve could be done on ORACLE in twenty minutes.

Householder and the Mathematics Panel used ORACLE to analyze radiation and shielding problems. In 1957, Hezz Stringfield and Ward Foster, both of the Budget Office, also adopted ORACLE for more mundane but equally important tasks—annual budgeting and monthly financial accounting. As one of the last "homemade computers," ORACLE became obsolete by the 1960s. The laboratory then purchased or leased its mainframe computers from commercial suppliers. From the initial applications of ORACLE to nuclear aircraft problems, computer enthusiasm spread like lightning throughout the laboratory, and in time, use of the machines became common in all the laboratory's divisions.

Particle Counting

Scintillation spectrometers and multichannel analyzers were other machines that benefited from—and contributed to—the laboratory's involvement with the nuclear aircraft project and its related studies of atomic particle behavior and radiation damage.

In 1947, German scientists observed that some crystals emitted flashes of light when struck by radiation beams and that the intensity of the flash was proportional to the radiation's energy. By 1950, a scientific team at ORNL led by P. R. Bell devised an improved scintillation spectrometer to measure the number and intensity of light flashes emanating from crystals exposed to radiation. Electronic recording of these measured flashes by multichannel analyzers permitted complete and rapid analysis of the particle and gamma ray energies.

Bell's group later converted the scintillation spectrometer into a medical pulse analyzer and developed a "scintiscanner" and an electronic probe to assist physicians using radioisotopes to locate tumors without surgery. In 1956, Bell's team received funding from AEC to continue this work, and they formed a Medical Instruments group in the laboratory's Thermonuclear Division at Y-12, where they primarily investigated fusion energy. Later, they incorporated electronic computers in medical scanners to improve diagnostic techniques. Commercial versions of the machines they invented became common at major medical centers throughout the world.

Radiation Damage

Laboratory research in the solid-state sciences received a boost from radiation damage studies conducted under the auspices of the nuclear aircraft project in the early 1950s. Prolonged exposure to radiation alters the properties of solids and often compromises their ability to serve as structural material in a reactor.

Under Douglas Billington, the Physics of Solids Institute was established in 1950 to pursue radiation damage and related solid-state investigations. Formed by joining the Solid State Section of the Physics Division with a section of the Metallurgy Division, the institute occupied a new laboratory building constructed south of the graphite reactor. Becoming the Solid State Division in 1952, this group quickly expanded scientific knowledge of radiation damage and other phenomena in solids.

"Inasmuch as a thorough understanding of the normal behavior of solids is necessary for a complete understanding of the effects induced by nuclear radiation in metals and other solids," Billington declared in 1950, "studies in related solid state fields are being carried out in conjunction with the radiation effects experiments." One notable discovery, made by Mark Robinson and Ordean Oen, was the prediction of the "ion channeling" phenomenon, in which charged particles move undisturbed for long distances between a solid's atomic layers, which number in the millions.

Investigations of radiation damage in connection with shielding and reactor development also became central to the early work of ORNL's Biology Division at the Y-12 plant. Biologists learned that nucleoproteins, present in living cell nuclei and essential to normal cell functioning, are highly sensitive to ionizing radiation. Paper chromatography and ion-exchange methods used to separate and analyze compounds, ORNL researchers reasoned, could help scientists and medical researchers measure and gauge this sensitivity.

After applying ion-exchange chromatography to the separation of fission products and starting the laboratory's radioisotopes program, Waldo Cohn used the same technique to separate and identify the constituents of nucleic acids. From this work came the discovery with Elliot Volkin that ribonucleic acid (RNA) has the same general structure as deoxyribonucleic acid (DNA), a concept that had a fundamental impact on molecular biology, virology, and genetics.

Of Mice and Mammals

By 1949, ORNL had ten thousand mice housed in renovated facilities at the Y-12 plant. Research on the mice, led by the Biology Division's William and Liane Russell, was designed to advance understanding of radiation effects on mammals. According to William Russell, mice are used for genetic studies because they have few diseases, are economical to feed and maintain, have a rapid reproduction rate, and have the same essential organs found in human beings.

Liane Russell's 1950 survey of the gestation period of mice, designed to examine their sensitivity to radiation, yielded valuable information about critical periods during embryo development. Russell and her colleagues showed that radiation-induced changes of cells were more likely to occur during gestation. Largely because of her research, women have been cautioned about X-ray examinations during pregnancies.

The Russells, a cosmopolitan husband-and-wife team from England and Austria, came to Oak Ridge in 1948, expecting to find a backward community with minimal social and cultural opportunities. The Biology Division had an international clientele, however, and Liane Russell was surprised by the extent to which the world beat a path to Oak Ridge and the laboratory. The Russells became renowned for taking their international guests on mountain hiking trails. They later played key roles in the creation of the Big South Fork National River Recreation Area, a wilderness preserve just north of Oak Ridge.

Technology School

Just as the Biology Division had an international reputation, the Oak Ridge School of Reactor Technology established in 1950 also enjoyed national prestige. Because reactor technology was security-sensitive and could not be taught in universities, AEC, with considerable support from Captain Rickover and the navy, sponsored this school for outstanding engineers and scientists. Frederick VonderLage, the school's first director, was a former navy officer who had taught physics at the Naval Academy. The faculty included ORNL staff, and the school's text consultant was the laboratory's Samuel Glasstone, who published several overviews of nuclear reactor technology.

The fifty members of the school's first class in 1950 came from AEC, government contractors, and the armed services; the second class came largely from industries needing trained personnel in reactor engineering and operations; later, college graduates planning to work in the nuclear industry were accepted.

Students took courses in reactor technology that covered reactor neutron physics, radiation damage, reactor materials, chemical separations processes, and experimental reactor engineering. They spent a year in Oak Ridge and supplemented their classroom training with part-time research assignments at ORNL. After two semesters, students would load fuel into the movable assembly in the bulk shielding "swimming pool" reactor, plotting the curve as fuel was added and the flux increased. They then compared the onset of critical mass with their predictions. Later, they spent a summer investigating specific problems, often analyzing a reactor design under consideration by AEC and then submitting a thesis on its feasibility.

The school expanded during the 1950s, occupying a new building completed by the laboratory in 1952 and specializing in advanced subjects not taught at universities. Under director Lewis Nelson, the school in 1957 joined six universities in offering a standard two-year curriculum. At the end of the decade, it enrolled its first international students. Five years later, the school closed when university science and engineering programs began to offer this type of specialized instruction. Of its nearly one thousand enrollees during the school's fifteen years of instruction, all but ten completed the course. Some of its graduates subsequently became leaders in the nuclear industry.

Flying High

When Union Carbide assumed management of ORNL, the graphite reactor was the only nuclear reactor on the Oak Ridge reservation. By 1953, the laboratory had three reactors operating, two nearing completion, and several others in various stages of planning and development.

In addition, it had high-speed computers, high-energy cyclotrons, and Van de Graaff particle accelerators. Equally important, the laboratory had succeeded in assembling an aggressive research staff that worked with a sense of urgency rivaling that of the war years.

As the laboratory expanded its reactor and shielding programs in response to the nuclear aircraft project and acquired the Y-12 research organization in the early 1950s, administrative realignment became necessary. Electronics experts from the Physics Division, for example, moved into an Instrumentation and Controls Division, and the shielding group became a separate Neutron Physics Division (later renamed the Engineering Physics Division, then the Engineering Physics and Mathematics Division). The Mathematics Section also became an independent division. Similar organizational changes took place in chemistry, reactor technology, and other ORNL research pursuits.

By 1953, laboratory personnel numbered 3,600, more than double the wartime peak; the staff was divided into fifteen research and operating divisions. "I am sometimes appalled by the size and scope of our operation here," Weinberg admitted privately to Wigner. "It seems that we have become willy-nilly victims, in a particularly devastating way, of the big operator malady."

In response, Wigner advised Weinberg to appoint a deputy and assistant directors to assist central management. Weinberg accepted the advice. John Swartout, director of the Chemistry Division, became Weinberg's assistant director in 1950 and deputy director in 1955. For administrative functions, Swartout became "Mr. Inside," while Weinberg was "Mr. Outside." Other assistant directors of the early 1950s included Elwood Shipley, Charles Winters, Robert Charpie, Walter Jordan, Mansell Ramsey, Ellison Taylor, and George Boyd.

"There is," observed Weinberg, "a hierarchy of responsibility in which management on each level depends on the integrity and sense of responsibility of the next level to do the job sensibly and well." This line of responsibility from individual to group leader to section chief to division director to assistant or associate director to laboratory director remained the prevailing administrative framework within ORNL during the ensuing decades.

The prime force behind laboratory expansion during the early 1950s ended in 1957, when Congress objected to continuing the costly nuclear aircraft project in the face of supersonic aircraft and ballistic missile development that made the nuclear aircraft concept unnecessary. In response to this congressional decision, the laboratory shelved its aircraft shielding and reactor prototype investigations. In 1961, President John Kennedy canceled the remainder of the nuclear aircraft project.

The scientific data gleaned for the aircraft project, however, soon proved useful when the laboratory undertook the design of a molten-salt reactor for electric power production. William Manley, a veteran of the nuclear aircraft program, later pointed out that the knowledge gained in handling liquid metals and fused salts also proved useful in the design of nuclear generators and reactors for use in space. As ORNL metallurgist George Adamson summarized it, "The program quite literally didn't get off the ground, but out of it grew the base for the high-temperature materials technology needed by NASA and in several industrial fields."

Although the nuclear aircraft project stalled, the laboratory's participation in efforts to apply nuclear energy to vehicle propulsion continued briefly in consultation with the Maritime Commission, which in 1957 built a nuclear-powered merchant ship. The twenty-one-thousand-ton ship propelled by a pressurized-water reactor was a floating laboratory, demonstrating the feasibility of commercial ships propelled by nuclear energy. At ORNL, a maritime reactors group headed by Alfred Boch provided technical review of the ship reactor design, while other ORNL units assisted with on-board health monitoring, environmental studies, and waste disposal. Completed in July 1959, the *N.S. Savannah* could remain at sea for three hundred thousand miles without refueling, proving the scientific and engineering feasibility of such ships. Nuclear-powered ships, however, could not compete economically with oil-fired vessels; thus, the *N.S. Savannah* became the first and last U.S. ship of its kind.

In the 1960s, the laboratory became involved in nuclear-power studies for the national space program, and in the 1980s it studied space reactors for the Strategic Defense Initiative. Despite these efforts, it is fair to say that the laboratory's work on the *N.S. Savannah* marked the end of its nuclear transportation programs. Postwar dreams of nuclear-powered trains, automobiles, aircraft, and tractors ended, but the scientific findings that evolved from these endeavors found surprising applications in other areas in the years ahead.

Chapter 4

Olympian Feats

A symbol of peaceful international competition in the ancient world, the Olympics were revived in modern times, not only in quadrennial athletic performances but in scientific competitions as well. Sparked in 1953 by President Dwight Eisenhower's call for international cooperation in the peaceful uses of atomic energy, scientists worldwide showcased their achievements in 1955 and 1958 at international conferences that resembled the athletic Olympics. In these competitions, the world-class research at ORNL often took the laurels.

Science during the 1950s became a full-blown instrument of foreign policy, both in Cold War weapons competition and in peaceful applications of nuclear science, especially nuclear fission reactors and fusion energy devices. As an international center for nuclear fission research, by the mid-1950s the laboratory had as many as six reactors under concurrent design and construction. The laboratory's chemical technology expertise also made it a leader in reactor fuel reprocessing and recovery.

Both these programs earned ORNL plenty of prestige at the 1955 scientific Olympics. And, in 1958, the laboratory's tiny fusion energy research effort vaulted above larger programs elsewhere to win the gold at the second international conference on peaceful uses of the atom.

The laboratory and other AEC facilities also ascended the ladder of experimental reactor development. In 1953, for example, the laboratory's experimental homogeneous reactor first generated electric power. Elsewhere, other nuclear mileposts were passed: For example, a demonstration atomic reactor to propel submarines and an experimental breeder reactor began operating in Idaho, and the first university research reactor was unveiled at North Carolina State University.

In a dramatic speech on the future of the atom to the United Nations in 1953, President Eisenhower pledged the United States "to find the way by which the miraculous inventiveness of man shall not be dedicated to his death, but consecrated to his life." The president's "Atoms for Peace" speech, hailed throughout the world as a prologue to a new chapter in the history of nuclear energy, was to guide the research efforts of AEC and ORNL for years to come. The initiative, Alvin Weinberg declared, would make nuclear science the "touchstone of peace."

Soon after his seminal address, President Eisenhower signed the 1954 Atomic Energy Act, which was designed to foster the cooperative development of commercial nuclear energy by AEC and private industry. In response, AEC began a massive declassification of nuclear science data for the benefit of private users. ORNL, in turn, assumed a key role in AEC's five-year plan to develop five new demonstration nuclear reactors.

Launched in 1954, the AEC plan called for the construction of a small pressurized-water reactor by Westinghouse Corporation; an experimental boiling-water reactor by Argonne National Laboratory; a fast breeder reactor, also by Argonne; a sodium graphite reactor by North American Aviation; and an aqueous homogeneous fuel reactor by ORNL.

Beyond its work on the homogeneous reactor, ORNL in the 1950s— as a national center for nuclear chemistry and chemical technology— focused on developing fluid fuels for nuclear reactors. In this experiment, the laboratory concentrated on three possible options: fuels in solution, fuels suspended in liquid (slurries), and molten-salt fuels.

Each of these options posed fundamental challenges in chemistry and chemical technology. Moving confidently from solids to liquids to gases in support of AEC efforts on behalf of the atom, the laboratory also conducted research for heterogeneous, solid-fuel reactors and drafted conceptual designs for a transportable army package reactor, a maritime reactor, and a gas-cooled reactor.

The Cold War and President Eisenhower's "Atoms for Peace" speech re-energized and refocused ORNL's research efforts. In effect, it gave the

Leo Holland describes the laboratory reactor to President Dwight Eisenhower at the 1955 Geneva conference.

laboratory a multifaceted research agenda, many aspects of which were tied to the development and application of nuclear power.

Summarizing the effect of the nation's postwar aims on the work of the laboratory, Director Clarence Larson commented, "1954 has witnessed the transition that many of us have hoped for since the war. The increasing emphasis on peacetime applications of atomic energy," he went on to say, "has been a particular source of gratification."

Aqueous Homogeneous Reactor

In addition to the nuclear aircraft reactor, the bulk shielding reactor, and the tower shielding facility built as part of its aircraft nuclear project for the air force, ORNL had three other major reactor designs in progress during the mid-1950s: its own new research reactor with a high neutron flux; a portable package reactor for the army; and the unique aqueous homogeneous reactor that combined fuel, moderator, and coolant in a single solution (designed as one of five demonstration reactors under AEC auspices).

Initial studies of homogeneous reactors took place toward the close of World War II. It pained chemists to see precisely fabricated solid-fuel elements of heterogeneous reactors eventually dissolved in acids for the removal of fission products, the "ashes" of a nuclear reaction. Chemical engineers hoped to design liquid-fuel reactors that would dispense with the costly destruction and reprocessing of solid-fuel elements. The formation of gas bubbles in liquid fuels and the corrosive attack of the high-temperature fuels on materials, however, presented daunting design and materials challenges.

With the additional help of experienced chemical engineers brought to ORNL after its acquisition of Y-12 laboratories in 1950, the laboratory proposed to address these design challenges. Rather than await theoretical solutions, laboratory staff attacked the problems empirically by building a small, cheap, experimental homogeneous reactor model.

A homogeneous (liquid-fuel) reactor had two major advantages over heterogeneous (solid-fuel and liquid coolant) reactors. Its fuel solution would circulate continuously from the reactor core through a processing plant that would remove unwanted fissionable material. Thus, unlike a solid-fuel reactor, a homogeneous reactor would not have to be taken off-line periodically to discard spent fuel. Equally important, a homogeneous reactor's fuel, and the solution in which it was dissolved, could serve as the source of power generation. For this reason, a homogeneous reactor held the promise of simplifying nuclear reactor designs by packaging the fuel and power source in one place.

A building to house the experimental homogeneous reactor was completed in March 1951. The first model for testing the feasibility of a homogeneous reactor used uranyl sulfate fuel. After leaks in the high-temperature pressure piping system were plugged, the power test run began in October 1952, and the design power level of one megawatt was attained in February 1953. The reactor's high-pressure steam twirled a small turbine that generated 150 kilowatts of electricity, an accomplishment that earned its operators the honorary title "Oak Ridge Power Company."

Marveling at the homogeneous reactor's smooth responsiveness to power demands, Weinberg found its initial operation thrilling. "Charley Winters at the steam throttle did everything, and during the course of the evening we electroplated several medallions and blew a steam whistle

Director Alvin Weinberg, *second from left*, and colleagues use the homogeneous reactor test to power an electric light.

with atomic steam," he exulted in a report to Wigner, asking him to bring von Neumann to see it.

Despite his enthusiasm, Weinberg found AEC's staff decidedly bearish on homogeneous reactors and, in a letter to Wigner, he speculated that the "boiler band wagon has developed so much pressure that everyone has climbed on it, pell mell." Weinberg surmised that AEC was committed to the development of solid-fuel liquid-cooled reactors and ORNL demonstrations of other reactor types—regardless of their success—were not likely to alter its course.

Nevertheless, the laboratory dismantled its first homogeneous model reactor in 1954 and obtained authority to build a large pilot plant with a "two-region" core tank. The aim was not only to produce economical electric power but to irradiate a thorium blanket surrounding the reactor, thereby producing fissionable uranium-233. If this pilot plant proved successful, the laboratory hoped to accomplish two major goals: to build a full-scale homogeneous reactor as a thorium "breeder" and to supply cheap electric power to the K-25 plant for enriching uranium.

Initial success stimulated international and private industrial interest, and in 1955, Westinghouse Corporation asked ORNL to study the
feasibility of building a full-scale homogeneous power breeder. British
and Dutch scientists studied similar reactors, and the Los Alamos Laboratory built a high-temperature homogeneous reactor using uranyl phosphate fluid fuel. If the laboratory's pilot plant operated successfully, staff
at Oak Ridge thought that homogeneous reactors could become the most
sought-after prototype in the intense worldwide competition to develop
an efficient commercial reactor. Proponents of solid-fuel reactors, the op-
tion of choice for many in AEC, would find themselves in the unenvi-
able position of playing catch-up.

Army Package Reactor

Similar initial success flowed from studies at the Oak Ridge School of
Reactor Technology, where a study group in 1952 proposed a compact,
transportable package reactor for generating steam and electric power
at military bases so remote that supplying them with bulky fossil fuels
was too difficult and costly.

AEC and the Army Corps of Engineers expressed a great deal of
interest in this concept, and in early 1953, ORNL management met with
Colonel James Lampert and Army Corps of Engineers staff to initiate planning for such a mobile reactor. Alfred Boch and staff in the Electronuclear
Division were given responsibility for designing this small reactor. They
selected a heterogeneous, pressurized-water, stainless steel system design
that could use many standard components for easy replacement at remote
bases. Walter Jordan led an ORNL team that drew up specifications for a
package reactor capable of generating ten megawatts of heat and two
megawatts of electricity. General Samuel Sturgis, chief of the Army Engineers, decided to build the reactor at Fort Belvoir, Virginia, where his
officers could be trained to operate it.

The package reactor was the first reactor built under bid by private
contractors. The Army Corps of Engineers, in fact, received eighteen
bids that ranged from $2.25 million to $6.9 million. The corps awarded
the contract to Alco Products (American Locomotive Company) in December 1954, and Alco completed the reactor in 1957.

With a core easily transportable in a C-47 airplane, the package reactor could generate power for two years without refueling, compared
to the fifty-four thousand barrels of diesel fuel that an oil-fired plant would
consume in the same time to produce the same amount of power. The
army later built similar package reactors for power and heat generation in the Arctic and other remote bases.

Purification

Ancient athletes considered the Olympics a purifying experience. Purification was also a preoccupation of scientists who participated in the
nuclear Olympics of the 1950s—not personal purification but fuel purification to enable nuclear reactors to operate more efficiently.

Although designers of the homogeneous reactor hoped to achieve
simultaneous reactor operation and fuel purification, other ORNL technologists led by M. D. Peterson, Frank Steahly, and Floyd Culler sought
improved methods of purifying spent fuels and recovering valuable
plutonium and uranium from once-used fuel elements in heterogeneous reactors.

The laboratory's interest in these efforts was reflected by the sub-
division of its Technical Division into the Reactor Technology Division
and the Chemical Technology Division in February 1950. The Reactor
Technology Division carried out laboratory responsibilities for reactor
development, while the Chemical Technology Division, following the
laboratory's "separations and recovery" experience during and after
World War II, sought to improve chemical separations processes.

The laboratory's most important achievement during World War II
had been the recovery of plutonium from graphite reactor fuel. Drawing on its wartime experience, the laboratory attained notable success
during the postwar years, recovering uranium stored in waste tanks
near the graphite reactor. The management at Hanford called on ORNL
staff to address similar recovery problems at its plutonium production
facilities in the state of Washington. ORNL also built a pilot plant to
improve Argonne National Laboratory's REDOX process for recovering plutonium and uranium through solvent extraction. The pilot plant
served as a prototype for an immense REDOX process plant completed

at Hanford in 1952. To recover uranium from fuel plates at AEC's Idaho reactor site, ORNL devised the so-called 25 process. A large plant using this process was completed there, also in 1952.

Recovery, separation, and extraction—the primary components of fuel purification—were big business at ORNL during the 1950s. Such efforts played a major role in developing the plutonium and uranium extraction (PUREX) process selected in 1950 for use at the Savannah River reactors. Two huge PUREX plants were built at Savannah River in 1954 and a third at Hanford in 1956. Later, large plants using the PUREX process were built in other nations, and some ORNL executives believe the PUREX process, in the end, may have constituted the laboratory's greatest contribution to nuclear energy.

By 1954, the laboratory's chemical technologists had completed a pilot plant demonstrating the THOREX process for separating thorium, protactinium, and uranium-233 from fission products and from each other. This process could isolate uranium-233 for weapons development and also for use in the proposed thorium breeder reactors.

During the 1950s, the laboratory's Chemical Technology Division served as AEC's center for pilot plant development, echoing the laboratory's wartime role in plutonium recovery and extraction. The succession of challenges it resolved—uranium-235 recovery, REDOX pilot plant, PUREX development, and THOREX pilot plant—swelled the ranks of the Chemical Technology Division from fewer than 100 people in 1950 to almost 200 in 1955. A similar doubling of personnel took place in the Analytical Chemistry Division; its staff increased from 110 people to 214 people during the same period.

The fuel purification program brought Eugene Wigner back to ORNL in 1954. Wigner had been working for Du Pont on the design of the Savannah River reactors when he agreed to return to Oak Ridge to apply his chemical engineering expertise to design a solvent extraction plant. Labeled "Project Hope" because it promised to extend the supply of fissionable materials for energy production, Wigner's 1954 study resulted in the design of a processing plant able to recover uranium-235 from spent fuel for reuse in reactors at a cost of $1 per gram, compared with the prevailing cost of $7.50 per gram of uranium from ore.

His study helped turn the attention of the laboratory's chemical technologists from improving individual processes for the recovery of uranium, plutonium, and thorium to developing an integrated plant capable of separating all nuclear materials at a single site. The proposed power reactor fuel reprocessing facility would have competed with private industry, however, and eventually AEC decided not to construct it.

Oak Ridge Research Reactor

In 1953, ORNL received AEC approval to build a new research reactor. The reactor design, blueprinted by Thomas Cole's team, combined features of the materials testing reactor and the swimming pool reactor. With a thermal power rating of twenty megawatts, its neutron flux—the critical research element—was exceeded only by the materials testing reactor in Idaho.

After several construction delays, the new Oak Ridge research reactor was completed and reached its design power in March 1958. A flexible, high-performance reactor with an easy-to-access core, it proved useful for physics and materials research, irradiations, and neutron beam studies. Physicists Cleland Johnson, Frances Pleasonton, and Arthur Snell performed the research reactor's first scientific experiments. They examined the relative directions of neutrino and electron emissions in the decay of helium-6, thereby clarifying beta decay interaction and guiding improvements of the recoil spectrometry technique pioneered by Snell and his colleagues.

During the reactor's thirty years of service, it supported many scientific advances and became a tourist attraction as well. An impressive structure, silhouetted by the blue glow of radiation emanating from the core within its protective pool, the Oak Ridge research reactor was admired in person by Senator John Kennedy, Representative Gerald Ford, and other noted and aspiring political figures. Thanks to relaxed security requirements in the wake of President Eisenhower's call for international cooperation, the reactor also attracted many foreign scientists and dignitaries, such as the queen of Greece and king of Jordan, who came to the laboratory on other business but could not pass up an opportunity to see one of the facility's most spectacular reactors.

Alvin Weinberg, *second from right*, shows ORNL research reactor controls to Jacqueline Kennedy, Senator John Kennedy, and Senator Albert Gore, Sr.

1955 Geneva Conference

The laboratory's new research reactor was being designed at the same time that plans were being made for the first United Nations Conference on Peaceful Uses of the Atom. That conference was scheduled for Geneva, Switzerland, in August 1955. Ostensibly a staid, professional scientific meeting organized in response to Eisenhower's "Atoms for Peace" initiative, in reality it was an extravagant science fair with exhibits from many nations emphasizing their scientific achievements.

Never before had the accomplishments of nuclear power been placed on such a public stage. And never before had scientists so openly presented their findings as symbols of national prowess. Just as the athletic Olympics in the post–World War II era emerged as peaceful arenas for venting Cold War animosities, the 1955 Geneva conference on the atom became a platform for comparing the relative strengths of science in capitalist and communist societies.

Because critical comparisons of the exhibits, especially those brought by the Soviets and the Americans, were expected, AEC asked its laboratories for spectacular exhibit concepts. At Oak Ridge, Thomas Cole's suggestion that AEC build and display a small nuclear reactor was welcomed.

In early 1955, Charles Winters picked up on Cole's suggestion and led an ORNL research team that designed and fabricated a scaled-down version of the materials test reactor, which operated at one-hundred kilowatts instead of thirty megawatts. It became the first reactor to use low-enriched uranium dioxide fuel. After testing, the reactor was disassembled and shipped by air from Knoxville to Geneva, where the ORNL team reassembled and tested it.

Designed, built, tested, transported to Geneva, and reassembled in only five months, it became the most spectacular display at the conference, admired by political dignitaries such as President Eisenhower as well as by the public and the media. The reactor and the twenty-eight scientific papers presented to the conference by staff members gave ORNL a claim to the laurels of the international competition.

Heralding the multifaceted applications of peaceful atomic power, the Geneva conference captured the public's imagination. After the conference, the American exhibit returned home for a triumphant national tour, minus its most eye-catching element. The Swiss government purchased Oak Ridge's model materials testing reactor to use at a research facility.

"Our Laboratory stands today as an institution of international reputation," exulted Alvin Weinberg, who became ORNL director shortly after the conference. "This we sense from our many distinguished foreign visitors," Weinberg continued, "from the numerous invitations which our staff receives to foreign meetings, and in the substantial part which we played at Geneva. But with international reputation," Weinberg cautioned, "comes international competition." And, as any Olympic champion will tell you, as difficult as it is to win the first gold medal, it is even more difficult to sustain a level of performance unequaled by others.

Gas-Cooled Reactor

International exchange brought the laboratory a new assignment from AEC: to explore gas-cooled reactor technology. Although U.S. studies of gas-cooled reactors waned with the termination of the Daniels Pile investigations in 1948, British scientists successfully designed and built several large gas-cooled reactors in the early 1950s. In 1956, Congress directed AEC to develop firsthand experience with gas-cooled, graphite-moderated reactors. In response, AEC turned to ORNL, which formed a study team headed by Robert Charpie. The work of this team led to evaluations of the comparative costs of nuclear power produced by gas-cooled and water-cooled reactors.

The laboratory's initial findings seemed promising. In 1957, AEC made the laboratory responsible for designing fuel elements for an experimental gas-cooled reactor to be constructed in Oak Ridge. By early 1958, the laboratory had completed a conceptual design for a helium-cooled, graphite-moderated reactor. Its core was to be uranium oxide clad in stainless steel, although a team led by Murray Rosenthal also studied graphite-coated fuel particles as alternative fuel elements.

With the cooperation of TVA, in 1959 AEC began construction of an experimental gas-cooled reactor on the bank of Clinch River near the laboratory. This reactor was to serve as a power-generating prototype. Eight test loops inside the reactor would allow ORNL scientists to test various fuel elements. Construction delays and increasing project costs, however, soon caused the test loops to be eliminated from the design.

Then, in 1966, AEC ordered the project stopped even though all construction on the reactor had been completed and its fuel elements had been manufactured and fully evaluated. The light-water reactor industry had advanced so rapidly that the Oak Ridge prototype could no longer serve the potential commercial purposes for which it was planned. Despite initial promise, the experimental gas-cooled reactor design had become obsolete before it was operational.

Molten-Salt Reactor Technology

The laboratory launched another innovative nuclear reactor design in 1956 when Herbert MacPherson headed a team investigating the application of molten-salt technology. The laboratory's aircraft reactor experiments during the early 1950s used molten (fused) uranium fluorides (salts) as reactor fuel.

Molten-salt fuel could function at high temperatures and low pressures in a liquid system that could be cleansed of fission ashes without stopping the reactor. Like other liquid nuclear fuels, however, molten salts were highly corrosive and posed significant materials challenges. MacPherson's group studied molten-salt fuels and materials in the test loops built for the aircraft reactor project, conducted cost studies of molten-salt reactors, and focused on identifying corrosion-resistant materials for use in such reactors.

When an AEC task force in 1959 identified molten salt as the most promising of the liquid-fuel reactor systems, AEC approved a molten-salt reactor experiment. By 1960, the laboratory was designing an experimental molten-salt reactor using graphite blocks as the moderator. A uranium-bearing fuel of molten fluorides circulated through metal tubes made of a nickel-molybdenum alloy, called Hastelloy N, which had been developed earlier at the laboratory for the aircraft reactor.

Molten-salt reactor experiments continued at the laboratory throughout the 1960s and into the early 1970s. Carlos Bamberger and colleagues devised a method of obtaining the element thorium by extracting it from the virtually inexhaustible supply of granite rocks found throughout the Earth's crust. When bombarded by neutrons in a molten-salt reactor, the thorium was converted to fissionable uranium-233, another nuclear fuel. The laboratory's experimental molten-salt reactor demonstrated its capability of using the thorium-to-uranium-233 fuel system in 1969. But, again, initial ORNL success would prove difficult to sustain and the molten-salt reactor would never prove safe enough to enter the commercial arena.

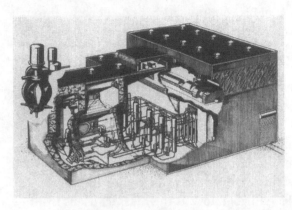

Drawing of the molten-salt reactor.

Project Sherwood

Alvin Weinberg described ORNL's use of the uranium-233 reactor fuel from thorium as "burning the rocks"; conversely, he called its secret investigations of producing fusion energy from heavy water (deuterium oxide), which could be obtained from seawater, "burning the sea." Thus, by the late 1950s, the laboratory's Olympians were searching for an inexhaustible energy supply extracted from either the Earth's crust or the seas. Using elements found in abundance in granite or seawater possibly would provide limitless energy.

The laboratory's fusion research efforts were no less Promethean than its fission research. Such research began in Oak Ridge in 1953 as a small part of AEC's classified Project Sherwood program. By the time of the second scientific Olympics at Geneva in 1958, the laboratory had become a world leader in fusion research.

Hydrogen nuclei release enormous energy when they fuse, as in a thermonuclear reaction associated with the detonation of a hydrogen bomb. Fusion temperatures of the hydrogen isotopes deuterium and tritium, however, are about 1 million degrees, a factor that presents obvious design challenges. In effect, fusion energy would prove too "hot" to handle.

Major research aimed at fusing these isotopes in a controlled thermonuclear reaction began in 1951, when Argentine President Juan Peron announced that scientists in his country had liberated energy through thermonuclear fusion without using uranium and under controlled conditions that could be replicated without causing a holocaust.

Peron's claim ultimately proved false, but it stimulated a host of international fusion research initiatives, including AEC's classified Project Sherwood. Legend has it that the name Sherwood emanated from the answer to the question, "Would you like to have cheap, non-polluting, and everlasting energy?" The answer was "Sure would." In reality, the name was derived from a complicated pun on the Sherwood Forest legend, which would test even the most vivid imagination. The pun involved robbing Hood Laboratory at the Massachusetts Institute of Technology to fund James Tuck's fusion research at Los Alamos.

To achieve fusion, scientists sought to contain a cloud, or plasma, of hydrogen ions at high temperature in a magnetic field. Because the plasma cooled if it touched the sides of its container, electromagnetic forces (pulling from different directions) were necessary to hold the plasma in the center away from the container's sides. If the plasma were suspended in the same place long enough and at a high enough density and temperature, scientists believed a fusion reaction would begin and become self-sustaining.

In its early years, Project Sherwood centered around three fusion devices. Princeton University had a stellarator, a hollow, twisted doughnut-shaped metal container, with electric wires coiled around it to supply a magnetic field and hold the charged hydrogen ions. Livermore Laboratory in California had a "mirror" device with a magnetic field stronger at its ends than in the middle to reflect hydrogen ions back to the middle of the field. And James Tuck's Perhapsatron at Los Alamos sought to contain the hot plasma through a "magnetic pinch"—that is, magnetic forces were designed to hold, or pinch, the plasma toward the middle of the container.

In Oak Ridge, the laboratory focused not on a particular device but on two problems basic to fusion devices: how to inject particles into the devices and how to heat the plasma to temperatures high enough to ignite the reaction.

With large surplus electromagnets on hand at Y-12 from the calutrons once used to separate uranium-235 from uranium-238, an ion source group in the Electronuclear Division—which included Ed Shipley, P. R. Bell,

Models of ORNL's direct current experiment (DCX) for
fusion research were displayed
at the 1958 Geneva conference.

Al Simon, and John Luce—became responsible for fusion research. Their background in electromagnetic separation and high-current cyclotrons led them to studies of energetic ion injection to create a hot plasma. Theoretical work showed promise, and in 1957, the laboratory formed its Thermonuclear Experimental Division with a staff of seventy people to pursue the fusion challenge. Personnel came from the Physics and Electronuclear divisions and from the discontinued aircraft reactor project.

In 1957, published stories and unsubstantiated rumors hinted that British scientists might have achieved a successful fusion reaction. Although exaggerated, the stories and rumors nevertheless encouraged greater emphasis on fusion research by both AEC and the laboratory. Moving a particle accelerator into Y-12 to provide a beam of high-energy deuterium molecular ions, Luce, Shipley, and their associates built the direct current experiment (DCX), a magnetic mirror fusion device. In August 1957, they "crossed the swords," injecting a deuterium molecular beam to a carbon arc that dissociated the beam into a visible ring of circulating deuterium ions (shaped like a bicycle tire with a slender circumference). This advance transformed Project Sherwood from a remote, abstract theory to a real but distant possibility.

Planning for a second Geneva conference on peaceful uses of the atom coincided with the laboratory's advance in fusion research. AEC Chairman Lewis Strauss, determined that the United States should achieve a triumph equal to that of 1955 at the 1958 scientific Olympics, threw AEC's full support behind fusion research. He hoped that American scientists could display an operating fusion energy device at the 1958 Geneva conference, just as they had displayed a successful nuclear reactor three years earlier.

"I have received a letter from Chairman Strauss exhorting the Laboratory to do everything it possibly can to have incontrovertible proof of a thermonuclear plasma by the time of Geneva," Weinberg informed laboratory staff. He went on to say:

> We are now engaged in this enterprise; we have mobilized people from every part of the Laboratory for this purpose and, with complete assurance of unlimited support from the Commission, we have put the work into the very highest gear. I can think of few things that would give any of us as much satisfaction as to have Oak Ridge the scene of the first successful demonstration of substantial amounts of controlled thermonuclear energy.

1958 Geneva Conference

By the time of the second Geneva conference on the peaceful uses of the atom in September 1958, intense media attention on the miracles of nuclear energy had somewhat jaded the public. Saturated for years with news about the potential miracles of nuclear energy but seeing few practical devices to illustrate its worth, Americans turned their attention to other matters. Moreover, Soviet scientists, so prominent at the 1955 conference, were no longer subjects of great public curiosity.

As a result of this diminishing public interest, the second Geneva conference turned out to be less a media circus and more a conventional scientific conference. In 1958, only schemes and devices for achieving controlled thermonuclear reaction through fusion enjoyed the glamor linked to the first conference.

The second conference, however, was then the largest international scientific conference ever held. Exhibits filled a huge hall built on the grounds of the Palais des Nations. Sixty-one nations participated, and twenty-one exhibited fusion devices, fission reactors, atom smashers, and models of nuclear power plants.

The United States, Great Britain, and the Soviet Union declassified their fusion research at the time of the conference, and Chairman Lewis Strauss resigned from AEC to lead the American delegation to Geneva. It took nearly ten hours to view the United States exhibit alone. The most popular attractions were models of ORNL's DCX fusion machine.

The laboratory provided two full-scale working models of its DCX machine to display its operating principles. Through viewing windows, visitors could see the beam, and the ring of ions wound around it like a ball of mystical yarn. Using a bit of showmanship, the laboratory made the trapped ring visible by dusting tungsten particles onto it from above.

Soviet fusion specialists took intense interest in the DCX display because they were also pursuing a molecular ion–injection approach to fusion. After the conference, other nations, drawing on the laboratory's experience, built DCX-type machines, making them fundamental tools for plasma research.

Yet, optimism over the future success of fusion energy soon faded. The supposed British achievement of fusion with a pinch-type device proved premature, and the ability of pinch machines ever to provide a stable plasma was questioned. Unstable plasma escaping the magnetic field also plagued the Princeton stellarator, and by the end of 1958, ORNL scientists learned that their carbon arc lost trapped ions, forcing the DCX staff to study different types of arcs and to plan an improved device, called DCX-2.

Alvin Weinberg, a proponent of nuclear fission and thorium breeding reactors, in 1959 compared Project Sherwood to "walking on planks over quicksand." Plasma physics was so novel then that the anticipated solid spots remained unknown; indeed, many scientists were skeptical that solid spots could exist under any conditions. "Working in this field requires a rugged constitution," Weinberg concluded, "but I'm told that those who can stand it find it stimulating."

Eugene Wigner reported that Soviet scientists were more cooperative at the 1958 Geneva conference than they had been in 1955, perhaps because the successful launching of the *Sputnik* satellite into orbit in 1957 had raised Soviet science to a new level of prominence. Wigner found them open about their nuclear fission and fusion energy research, but unwilling to share information about their space missions or their particle acceleration program. "Pure science in the Soviet Union still seems to be far from an open book," he observed.

Early Soviet achievements in space exploration sent shock waves throughout American political and scientific circles. Following the Soviet's successful launch of *Sputnik*, international scientific competition shifted from fission and fusion energy research to the race for space. As international scientific interests shifted, so did the focus of the federal government from AEC to the new National Aeronautics and Space Administration (NASA). Nuclear research remained an important aspect of America's scientific agenda, but it now had to share the policy spotlight with space issues. Geneva conferences on the atom were held occasionally after 1958, but none ever gripped the public imagination as had the first and second.

After the Gold

Nuclear reactor development at ORNL reached a pinnacle in 1956 and began a slow descent in 1957 with cancellation of its aircraft reactor program and troubles with its second experimental homogeneous reactor. In 1956, when the laboratory budget was $60 million and its staff reached 4,369, Weinberg boasted, "We are the largest nuclear energy laboratory in the United States, and we are among the half dozen largest technical institutions in the world."

With cancellation of the aircraft reactor in September 1957, the laboratory budget was slashed by 20 percent and its staffing cut to 3,943. About 1,500 personnel were at work on the aircraft reactor program. The 1957 reduction would have been even deeper if the laboratory had not absorbed some people into the molten-salt reactor, gas-cooled reactor, and Sherwood fusion programs.

Moreover, the Eisenhower administration froze the laboratory's budget in 1957, forcing postponement of a major building expansion program that included an east wing of the general research building, an instruments building, and a metallurgy and ceramics building, which together would have added a half-million square feet of work space. Weinberg called these actions "cataclysmic setbacks" that ranked with the loss of the materials testing reactor in 1947.

Homogeneous Reactor II

After successful operation of the first aqueous homogeneous reactor in 1954, the laboratory proceeded with the design of a larger homogeneous reactor on a pilot-plant scale. Whereas the first reactor had been a onetime experiment to prove yet unproven theoretical principles, the second reactor, sometimes identified as the homogeneous reactor test, was designed to operate routinely for lengthy periods.

The second homogeneous reactor was fueled by a uranyl sulfate solution containing ten grams of enriched uranium per kilogram of heavy water, which circulated through its core at the rate of four hundred gallons per minute. Its fuel loop included the central core, a pressurizer, separator, steam generator, circulating pump, and interconnected piping. Its core vessel was made of zircaloy, which was approximately a yard in diameter and centered inside a sixty-inch spherical pressure vessel made of stainless steel. A reflector blanket of heavy water filled the space between the two vessels.

Perhaps the most exotic nuclear reactor ever built, it gave laboratory staffers trouble from the start. During its shakedown run with pressurized water, chloride ions contaminated the leak detector lines, forcing the replacement of that system and delaying the power test by six months. And this marked just the beginning of the laboratory's "homogeneous" woes.

The homogeneous reactor had run many megawatt hours at below peak power from January into February 1958 when it became apparent that its outside stainless steel tank was corroding too rapidly. In April the reactor reached its design power of five megawatts, but in September a hole suddenly formed in the interior zircaloy tank. Viewing the hole through jury-rigged periscope and mirrors, operators determined that the hole had been melted into the tank, meaning that uranium had settled out of the fuel solution and lodged on the tank's side.

By the end of 1958, AEC considered abandoning the homogeneous reactor, and Eugene Wigner came to ORNL to inspect it personally. "The trouble seems to be that the rich phase absorbs to the walls and forms a solid layer there," Wigner told AEC staff. He thought altering the flow of fluid through the core would provide a velocity needed to prevent settlement of the uranium on the tank walls. "It is my opinion that abandoning the program would be a monumental mistake," he warned, pointing out that the reactor could convert thorium into uranium-233 to supplement a dwindling supply of uranium-235.

AEC allowed the laboratory to alter the reactor flow and continue its testing in 1959. These activities were accomplished by interchanging the inlet and outlet to reverse the fluid flow through the reactor. Several lengthy test runs followed during 1959, and the reactor operated continuously for 105 days—at the time, a record for uninterrupted operation of reactors. The lengthy test run demonstrated the advantages of a homogeneous system, where new fuel could be added and fission products removed while the reactor continued to operate.

Near the end of the year, however, a second hole burned in the core tank. Laboratory staff again patched the hole through some difficult remote dentistry and started another test run, but making the homogeneous reactor run reliably and safely proved challenging. In view of these unresolved difficulties, Pennsylvania Power and Light Company and Westinghouse Corporation abandoned their proposal to build a homogeneous reactor as a central power station.

During the shutdown and repairs, Congress viewed the homogeneous reactor troubles unfavorably, and in December 1960, AEC directed the laboratory to end testing and turn its attention to molten-salt reactor and thorium breeder development. The last homogeneous reactor test run continued until early 1961. For months, the reactor operated at full power until a plug installed earlier to patch one of the uranium holes disintegrated.

Ecological Challenges

Even as the laboratory climbed the acropolis of nuclear energy, challenges relating to nuclear fission and the laboratory's missions arose.

The threat of radioactive fallout from atmospheric testing of nuclear bombs and the need to deal more effectively with ground-based hazardous wastes called for research by ORNL's scientists. The need to broaden the laboratory's research agenda yet avoid competition with private industry also challenged its management.

Until 1963, fission and fusion bomb tests were conducted in the atmosphere, causing deep public concern about radioactive fallout. A principal concern during the early 1950s was the fallout of strontium-90, a bone-seeking fission product that fell from windblown clouds to the soil, where it could be taken up by vegetation and eaten by cows, eventually winding up in milk consumed by humans.

To study this problem and other issues of radiation ecology, the laboratory, at the recommendation of Edward Struxness, hired Orlando Park, an ecologist from Northwestern University, as a consultant in 1953. The laboratory subsequently asked Park's student, Stanley Auerbach, to join its Health Physics Division.

Both Park and Auerbach were expert investigators of the effects of radioactivity on ecological systems, particularly how radioactive nuclides migrate from water and soil to plants, animals, and humans. A major issue in the early 1950s was how quickly strontium-90 in the soil was taken up by plants. In fact, this and other questions about radioactive fallout became issues in the 1956 presidential election, when citizens' concerns about the potential impact of radiation on public health heated up the Eisenhower–Adlai Stevenson campaign trail. During the same year, the laboratory expanded its scientific studies of radioactive fallout into the Radiation Ecology Section in the Health Physics Division with Auerbach as section chief.

Auerbach and his colleagues found a ready field laboratory for their work in the bed of White Oak Lake, a drained reservoir where ORNL once had flushed low-level wastes. Examining the native plants and even planting corn in the radioactive lake bed, the ecologists studied the manner in which vegetation absorbed nuclides from the environment. Investigations of insects, fish, mammals, and other creatures followed, enabling ORNL ecologists to establish international reputations in aquatic and terrestrial radioecology.

Taking advantage of the laboratory's isotopes, the ecologists used radioactive tracers to follow the movements of animals, the route of

Stanley Auerbach of the Environmental Sciences Division tests a vehicle in a forest near the laboratory.

chemicals through the food chain, and the rates of decomposition in forest detritus. Sponsoring national symposia on ecosystems and related subjects, their work added a great deal to the study of radioecology, an emerging scientific field that counted Auerbach and his colleagues among its founders. When atmospheric atomic bomb testing ended in 1963 and interest in fallout waned, the ecologists expanded their studies to nonnuclear matters, forming the nucleus of the Environmental Sciences Division, established at the laboratory in 1970.

The increasing number of nuclear reactors during the 1950s, both at the laboratory and throughout the nation, produced increasing volumes of radioactive waste and mounting concern about its disposal. In 1948, ORNL formed a Waste Disposal Research Section in the Health Physics Division, and in 1952, it completed a radioactive waste research laboratory building for waste-management studies supported by AEC and the national civil defense agency.

During World War II, the laboratory stored its radioactive wastes in underground tanks for later recovery of the uranium and released its low-level wastes untreated into White Oak Lake. To reduce the level of radioactivity entering White Oak Creek and eventually the Clinch River, the laboratory built a waste-treatment plant during the 1950s to remove strontium and other fission products from its drainage. The laboratory also sought to recover uranium and other materials from its underground tanks and pump the remaining wastes into disposal pits.

In 1953, the laboratory initiated a multipronged remediation program designed to address its higher-level waste-disposal problems. The Chemical Technology Division devised a pot calcination strategy that heated high-level liquid wastes in steel pots, converting the wastes into ceramic material for easier handling and storage. The Health Physics Division, under the direction of Edward Struxness and Wallace de Laguna, explored the hydrofracture disposal method used by the petroleum industry. The strategy called for drilling deep wells, applying pressure to fracture the rock substrata, and pumping cement grout mixed with radioactive wastes down the wells to spread into the rock cracks and harden.

Struxness also joined Frank Parker of the Chemical Technology Division in studies of waste disposal in salt mines, and in 1959, the laboratory tested this method by storing nonradioactive wastes in a Kansas salt mine. Such strategies seemed promising during the 1950s, but each presented difficulties and none permanently resolved the disposal challenges presented by radioactive materials.

As the laboratory's operating nuclear reactors increased in number and its fuel-processing program burgeoned, the safety of equipment and the health of its personnel became a growing concern. Such concerns came to the forefront after a serious nuclear mishap in England during the late 1950s.

At Windscale, England, a British graphite reactor caught fire in 1957 when its operators sought to anneal the radioactive debris stored in the graphite as a result of the "Wigner disease." (Annealing is a process of heating and slow cooling to increase a material's toughness and reduce its brittleness.)

Herbert MacPherson and an ORNL team visited Windscale to review

In early years, the laboratory buried low-level wastes such as discarded protective clothing in trenches.

the accident and consider its implications for operation of the laboratory's own graphite reactor. MacPherson reported the laboratory's reactor operated at lower power and higher temperature than the Windscale reactor and that a similar accident could not occur in Oak Ridge because the materials in the reactor walls were not subject to the same stress and therefore would not require the same degree of heating and cooling to anneal. In the early 1960s, in fact, the laboratory's graphite reactor was annealed three times without difficulties by reversing its airflow and slowly raising power.

Although no accidents involving reactors occurred at the laboratory, three threatening situations involving radioactive materials did take place in 1959. First, fission products accidentally entered the liquid waste-disposal system from the THOREX pilot plant and were trapped in a settling basin. Second, ruthenium oxide trapped on the brick smokestack's rusty duct work shook loose during maintenance, forcing the installation of more filters and scrubbers in the stack. And, third, a chemical explosion in the THOREX pilot plant during decontamination released about six-tenths of a gram of plutonium from a hot cell, spreading it onto a street and the graphite reactor next to the plant.

It was largely by chance that no personnel suffered overexposure from these accidents, and the laboratory immediately stopped its radiochemical operations for safety review. Improved containment measures followed, and Frank Bruce took charge of the laboratory's radiation safety and control office to implement stricter safety precautions. P. R. Bell, Casimir Borkowski, and colleagues also devised compact radiation monitors. One called the "pocket screamer" was worn in the pocket and chirped and flashed at a speed proportional to gamma dosage rate. These devices were supplied to laboratory personnel.

In addition to these challenges, ORNL found it increasingly difficult to keep background radiation at acceptable levels because the amount of radioactivity handled by the laboratory increased during the 1950s, while government regulations steadily reduced the permissible levels to which workers could be exposed.

Karl Morgan at ORNL and other health physicists throughout the nation maintained that the maximum permissible levels should be so low that hazards resulting from radiation would be no greater than other occupational hazards. Laboratory biologists, however, had obtained differing results in studies of the effects of background radiation. Arthur Upton, for example, found that mice subjected to low-level chronic radiation seemed to have an improved survival rate from infections or other biological crises. The issue of acceptable radiation levels thus would prove contentious not only in the public arena but within ORNL as well.

Competitive Challenges

Not only did the laboratory face international competition during the late 1950s, it increasingly encountered competition at home from private nuclear companies. By 1959, for example, the rapidly growing nuclear industry questioned the role of national laboratories, urging that some of their work be contracted to private industry or even that the laboratories be closed.

Partly as a result of these pressures, AEC circumscribed ORNL programs in the late 1950s. The agency, for instance, canceled the power reactor fuel-reprocessing facility that the Chemical Technology Division hoped to build in Oak Ridge. In 1959, the laboratory also recognized that it would soon lose its homogeneous and gas-cooled reactor programs.

In response to the expected decline in its nuclear reactor and chemical-reprocessing programs, the laboratory conducted an advanced technologies seminar in 1959 to identify possible missions beyond nuclear energy. The seminar recommended additional study of nationally valuable research programs that had not been commercially exploited. Desalination of seawater, oceanography, space technology, chemical contamination management, and large-scale biology were mentioned as potential broad avenues of inquiry.

While convinced that federal investment in national laboratories was too great to permit their abandonment, Weinberg recognized that a realignment of their missions was in order. Asked to forecast the role of science and national laboratories during the 1960s, Weinberg expressed his hope that they "will be able to move more strongly toward those issues, primarily in the biological sciences, which bear directly upon the welfare of mankind."

The Olympics of antiquity had begun as a single event: a long-distance race between the best runners of competing Greek city-states. The modern Olympics, in the post–World War II era, have been transformed into a carnival of sporting events in which athletes worldwide display their diverse athletic skills as runners, swimmers, equestrians, weight lifters, skeet shooters, and volleyball and basketball players.

In the same way, the scientific Olympics in which ORNL competed in the 1950s began as a contest measuring the scientific prowess of the Soviet Union and the United States. The laboratory, as one of America's primary institutions for scientific research, had a simple goal: display the nation's scientific talent and accomplishments in the most dramatic way possible.

By the early the 1960s, however, the contest became more diverse and complicated. Space issues eclipsed the importance of nuclear research as the most important symbol of a nation's scientific capabilities; other goals began to compete for the laboratory's resources and energies; and the initial successes of fission and fusion research proved difficult to replicate.

In short, like Olympic runners who followed in the path of their earliest brethren, ORNL scientists by the end of the 1950s found they would have to share the arena with other figures and other events. As the laboratory entered the 1960s, its work would be less dramatic but no less important, and its focus more diverse but no less compelling.

Chapter 5

Balancing Act

In 1961, Director Alvin Weinberg predicted historians would view atom-smashing accelerators, fission reactors, and fusion energy machines as prime symbols of modern history, just as the Egyptian pyramids and Roman Colosseum have come to symbolize those ancient cultures. The same year Weinberg made that prediction, however, ORNL activities began to shift slowly from a reliance on nuclear science and engineering hardware to sciences related to environmental restoration, nonnuclear energy, and social engineering.

In the 1960s, the call by congressional committees for AEC to expand and diversify national laboratory programs to create more "balanced laboratories" struck a responsive chord in Oak Ridge. Program disruptions, which followed the terminations of the materials test reactor in 1947, the aircraft nuclear reactor in 1957, and the homogeneous reactor test in 1961, taught laboratory management the dangers of relying on a few large hardware programs. In addition, national participation in the space race intensified competition for federal research dollars.

Responding to the "balanced laboratory" challenge, Director Weinberg organized an advanced technologies seminar to consider the laboratory's future. "What we should try to do is to identify long-range, valid missions which in scope and importance are suitable for prosecution by ORNL," he said. "Most missions of this sort will probably not fall in the field of nuclear energy," Weinberg added. "This need not bother us since in the long run," he predicted, "ORNL very possibly will not be in nuclear energy exclusively."

As a member of science panels advising Presidents Dwight Eisenhower and John Kennedy, Weinberg aggressively sought to use ORNL expertise to help solve national and international environmental and social problems. Under Weinberg's leadership, and the leadership of Alexander Hollaender in biology, the laboratory broadened its programs during the 1960s. Although basic nuclear science continued as a mainstay, the laboratory increasingly focused on the applications and safety of nuclear energy: how, for example, commercial nuclear power could reduce air pollution and chemical contamination resulting from burning fossil fuels, or produce fresh water from the seas for agricultural and industrial applications.

The laboratory had been a nuclear science research center from its inception; in 1961, it took the first steps toward becoming a national laboratory in a broader sense. Before 1961, all ORNL funding came from AEC. A decade later, about 14 percent of its hundred-million-dollar annual budget came from agencies outside AEC, usually for programs connected with civil defense, desalination, space travel, and cancer research.

Information, Please

An immediate local result of Weinberg's service on presidential science panels was the implementation of programs to manage the scientific "information revolution."

A historian in 1961 pointed out that the first science journal was published in 1665; the number climbed to one hundred in 1800, ten thousand in 1900, and forty thousand by 1961. Science was being buried under a blizzard of new publications. This information explosion, atop increasing specialization and a threatened shortage of scientists, the historian predicted, could cause the collapse of science by 1970. Placed in charge of a presidential task force investigating this ominous trend, Weinberg echoed the historian's sentiments when he said scientists were "being snowed by a mound of undigested reports, papers, meetings, and books."

To help solve this crisis, Weinberg proposed the creation of "information centers." Instead of traditional libraries with stacks of books and

The laboratory's desalination pilot plant used efficient heat-transfer tubes to desalt sea water.

shelves of journals available to researchers, Weinberg's centers would consist of scientists who would read everything published in their specialty, review the data, and provide their colleagues with abstracts, critical reviews, and bibliographic tools. These scientific "middlemen" would contribute to science directly by perceiving new intellectual ties during their in-depth reviews of the literature and by applying their new perceptions to their own research.

Weinberg's recommendation received broad acceptance. Nationally, more than three hundred science information centers were formed, including a dozen at the laboratory. Among the places designated as an ORNL center was the nuclear data project, begun at the laboratory in the 1940s by Kay Way. In 1949, Way moved the nuclear data project to Washington, D.C., under sponsorship of the Bureau of Standards and later the National Academy of Sciences. Weinberg brought Way and her team of seven physicists back to ORNL in 1964, where they continued the systematic collection and evaluation of nuclear data, publishing it in tabulated form for use by researchers.

Other ORNL information centers specialized in the fields of accelerators, atomic-collision cross-sections, charged particles, engineering, isotopes, nuclear safety, materials research, shielding, and environmental and life sciences. Coordinated by Walter Jordan and François Kertesz, the centers disseminated the information they collected largely by publishing review journals, annotated bibliographies, charts, and digital computer codes. Widely acclaimed, many of these publications and services continued to inform scientists into the 1990s.

Desalting the Seas

Although less successful in the long run than the information centers, the laboratory's research into desalting seawater attracted the most public and political attention of all its endeavors to achieve "balance." The program had two distinct points of origin.

As a result of its research into fluid-fuel reactors and the chemical processing of nuclear fuels, the laboratory employed some of the world's foremost solution chemists. Some of these chemists had become intrigued by the chemistry of desalting seawater, and they voiced support for desalination as a new laboratory mission in Weinberg's advanced technology seminars. A committee headed by Richard Lyon subsequently explored its potential with the Office of Saline Water, a research arm of the Department of the Interior.

In Washington, D.C., Weinberg discussed desalination with other members of the presidential science panel, especially Secretary of the Interior Stuart Udall's science advisor. Managers at the Interior Department's Office of Saline Water lacked enthusiasm for funding desalting research at ORNL, but Udall and Glenn Seaborg, chairman of AEC, orchestrated a "shotgun wedding" between the two federal agencies.

Funded initially at six hundred thousand dollars per year by the Office of Saline Water and AEC, a team of twenty solution chemists and engineers led by Kurt Kraus investigated the physical chemistry of seawater, focusing on hyperfiltration (reverse osmosis) to remove salts and contaminants from water. Development of dynamic membranes for rapid production of fresh water from the seas earned the team wide recognition.

A second phase of the laboratory's desalting work originated with

Philip Hammond, brought to the laboratory from Los Alamos National Laboratory in 1961. Hammond's maxim was "Bigger is cheaper." He contended that large nuclear reactors could produce power and heat cheaply enough to desalt seawater, providing fresh water for agriculture and electric power for industry. Although skeptical at first, laboratory management eventually found Hammond's concept to have merit, a belief also expressed by an independent task force of the Department of the Interior.

Presidents John Kennedy and Lyndon Johnson judged desalination to be in the national interest. Johnson, in fact, sought to make it an instrument of foreign policy, hoping to build nuclear desalination centers in arid regions such as the Middle East to reduce international competition for natural resources. Echoing the president, Weinberg said, "I can think of few major technical achievements, including manned exploration of space, that would have as much beneficial political impact as would making the deserts bloom with nuclear energy."

At the 1964 United Nations conference on peaceful uses of the atom in Geneva, President Johnson, Soviet Premier Nikita Khrushchev, and United Nations Secretary-General U Thant viewed the laboratory's proposed nuclear agro-industrial complexes favorably. Dubbed *nuplexes* by the media, these blueprints called for huge nuclear reactors to produce fresh water from the ocean for irrigating crops and for generating electric power for industry, thus fostering environments that balanced agriculture with manufacturing.

With international support, ORNL staff in 1964 traveled to Israel, India, Puerto Rico, Pakistan, Mexico, and the Soviet Union to assist with plans for desalination plants. In California, water-starved but people-packed Los Angeles laid plans to build a large desalination nuplex on an island off the coast. In private, however, Weinberg warned AEC chairman Seaborg that desalination publicity had outrun the technical capabilities, and that the laboratory needed increased research funding "so that the technical basis for the politicians' speeches always remains as firm as possible."

By 1965, when President Johnson announced his "Water for Peace" program, the laboratory had a hundred scientists studying desalination. Its water research team was developing evaporator tubes four times more efficient than earlier models at producing fresh water from the sea. In addition, the Rockefeller Foundation, which funded research

Gerald Goldstein tests a
GeMSAEC analyzer devel-
oped at the laboratory.

into disease- and drought-resistant seedlings to nurture the Green Revolution, became interested in nuplexes as potential food factories in pov-
erty-stricken nations. Former President Eisenhower and former AEC
chairman Strauss endorsed a desalination plant in the Middle East sponsored by private funds funneled through the International Atomic Energy Agency.

"In one sense it is premature to try to define the future role of
nuclear desalinization for agriculture, when no large city supply plant
is yet operating," warned Philip Hammond in 1966. "So far one plant
is under construction (in the Soviet Union), the Israeli plant has been
found feasible, and the MWD station (in Los Angeles) has reached the
final stages of negotiation. These pioneer plants are essential steps in
development of a brand new resource."

The desalination bubble burst as quickly as it had formed. In 1968,
Los Angeles abandoned its plans for a 150-million-gallon-per-day nuclear
desalination plant. The costs of nuclear plants had escalated so rapidly that
the plant no longer seemed economically feasible. As nuclear power costs

skyrocketed and the country's social and environmental concerns moved to the forefront, the media and political leaders lost interest in nuplexes. None was ever built, and funding for desalination research dried up.

"Solving today's social and economic problems with tomorrow's technology is risky," Weinberg lamented near the close of this effort by ORNL to become more "balanced." Yet the information obtained from desalination research later proved valuable for laboratory development of technologies to treat contaminated water and sewage. Plans to make the desert bloom had been pared to efforts to make water clean—no doubt a less glamorous initiative but nevertheless one with broad social and environmental purpose.

Big Biology

Alexander Hollaender's Biology Division prospered enormously during the laboratory's efforts to "balance" its research programs. Staffed by experts who studied the genetic and physical effects of radiation on living organisms, the division also hoped to shed light on the impact of radiation on the environment.

When Rachel Carson's *Silent Spring* was published in 1962, it stimulated intense public concern about the role chemical agents played in biological and environmental degradation. This widespread worry prompted increased research funding for the National Institutes of Health (NIH), whose managers soon received visits from Hollaender, Weinberg, and other ORNL staff. The discussions—and subsequent funding proposals—bore fruit during the 1960s in increased biological understanding and improved tools for science and medicine.

With support from the National Cancer Institute, the Biology Division opened a Biophysical Separations Laboratory, taking advantage of centrifuge designs by Paul Vanstrum and fellow researchers at the K-25 plant. The K-25 team had devised improved centrifuges to produce enriched uranium, and in 1961, a biology team headed by Norman Anderson, with advice from Jonas Salk of polio vaccine fame, adopted centrifuge technology to separate viruses from human leukemic plasma, hoping to identify a cure for leukemia. This striking use of nuclear separations technology to advance science and medical research led in several directions.

A hollow cylinder subdivided into sectors, which creates a zonal centrifuge whirling at high speeds, can separate substances at the molecular level into their constituents according to size and density. Anderson and his team experimented with centrifuges whirling up to 141,000 revolutions per minute. They learned the machines could separate impurities from the viruses causing polio and the Hong Kong flu. This finding had practical applications in large-scale separations required to produce vaccines against such diseases. By cleansing vaccines of foreign proteins, the zonal centrifuge could minimize the fever reactions that often accompanied immunizations. By the late 1960s, millions of people received vaccines that had been purified in zonal centrifuges, which also provided pure rabies vaccines for their pets.

Farmers and ranchers benefited from the research of Peter Mazur and Stanley Leibo of the Biology Division. They pioneered the freezing and transplanting of embryos of black mice into white mice in 1972, an early gene technology that spurred a revolution in animal husbandry and improvements in the quality and quantity of meat.

In other applications jointly sponsored by AEC and NIH, the molecular anatomy (MAN) program managed by Norman Anderson sought to identify the metabolic profiles and chemical characteristics of all cell constituents. Charles Scott and associates in the MAN program devised portable centrifugal analyzers commonly used later in medical clinics across the nation. Spinning at high speeds, these analyzers could separate and assay components of blood, urine, and other body fluids in minutes, recording the data on computers for medical diagnosis.

The best known of these machines was the laboratory's GeMSAEC, so named because its development was funded jointly by NIH's General Medical Sciences Division and AEC. Using a rotor spinning fifteen transparent tubes past a light beam, GeMSAEC displayed the results on an oscilloscope and fed the data into a computer, completing fifteen medical analyses in the time it previously took to perform one.

Another eye-catching development in the Biology Division emanated from the laboratory's search for powerful microscopes able to view and photograph objects the size of a few atoms. After the laboratory built an experimental microscope with high resolution in 1967, Oscar Miller and Barbara Beatty of the Biology Division placed frog eggs under the microscope and photographed genes in the act of making RNA. "I never expected to see the thread of life, the mysterious stuff that poets conjured long ago to explain the passage of the heartbeat from generation to generation across the eons," mused John Lear of the *Saturday Review of Literature*, who came from New York to peep into the microscope. "Yet today the thread lies clearly visible before me, under the lens of an electron microscope, here in the Tennessee hills."

In addition to funding from NIH for centrifuge and microscope research, the Biology Division received support in 1965 from the National Cancer Institute for a co-carcinogenesis research laboratory to investigate the complex biochemical events leading to cancer growth. This work took advantage of the 250,000 mice on hand in the Biology Division at the Y-12 complex.

Arthur Upton and his associates used the mice to study the physical effects of radiation and chemical agents on the environment and on human health. The experiments largely concerned airborne carcinogenesis, or the induction of lung cancer by exposure to pesticides, sulfur dioxide, urban smog, or cigarette smoke, both singly and together. Mice exposed to these irritants in an inhalation chamber were then raised in a clean environment while scientists observed the formation of tumors. Upton later left ORNL to become director of the National Cancer Institute.

At the time, the potentially dangerous components of cigarette smoke were largely unknown. To overcome this handicap, a lung cancer task force from the Analytical Chemistry Division became involved in carcinogenesis studies when it devised "ORNL Smoking Machine, Model Number 1," which smoked six cigarettes at a time, even mimicking human drags on the weed. "This isn't an easy task by any means," commented Herman Holsopple, who built the machine. "Every component in cigarette smoke must first be identified and then studied for its biological effect on humans, and right now we're just trying to identify some of the components."

To assess how environmental hazards threaten human health required big protocols, large epidemiological studies, and expensive machines—just the requirements that Big Biology at the laboratory could provide. By the late 1960s, the Biology Division, employing 450 personnel, had become the largest division in the laboratory.

Medical knowledge and clinical machines developed at the laboratory with NIH funding stimulated the formation of a University of Tennessee–Oak Ridge National Laboratory Graduate School of Biomedical Science. Thanks to grants from the Ford Foundation, the laboratory had entered a cooperative program with the University of Tennessee during the early 1960s. As many as fifty ORNL scientists worked several days each week as laboratory researchers and spent the remainder of the week as members of the university science faculty.

This cooperation laid the groundwork for a challenge presented in 1965 by James Shannon, director of NIH. Shannon planned a graduate school in biomedical science near NIH headquarters at Bethesda, Maryland, and as a condition for expanding NIH programs at ORNL, he urged the creation of a similar graduate school in Oak Ridge.

After Weinberg, Clarence Larson, and James Liverman obtained approval for such a school from the AEC commissioners and Donald Hornig, President Johnson's science advisor, Weinberg asked Andrew Holt, president of the University of Tennessee, if he would be interested in developing the school cooperatively. "Our location in Appalachia and the strong contribution which a major new biomedical program would make to President Johnson's Great Society," Weinberg told Holt, "should enlist the aid of our U.S. Senators and Congressmen as well as the President."

President Holt and university trustees approved the school in late 1965. Governor Frank Clement contributed one hundred thousand dollars of state funds, and Clarence Larson arranged a hundred-thousand-dollar contribution from Union Carbide. In 1967, the UT-ORNL Graduate School of Biomedical Science opened, with Clinton Fuller as its first director. It was staffed chiefly by Biology Division personnel holding joint appointments with the University of Tennessee and ORNL.

Civil Defense

At the same time the Graduate School of Biomedical Science was being organized, Weinberg explored the formation of a Civil Defense Institute at Oak Ridge. The origins of this concept may be traced to the closing ceremony for the laboratory's historic graphite reactor in November 1963.

AEC Chairman Seaborg, Eugene Wigner, Richard Doan, and other alumni of the laboratory's wartime "atomic" campaign returned to Oak Ridge for a nostalgic ceremony formally deactivating the graphite reactor on November 4, 1963, after twenty years of service. The next morning, Wigner learned that he would receive the Nobel Prize in physics, an award adding to his public visibility and prominence. At the time, he was campaigning for improved national civil defense. "According to the preamble to the Constitution, one of the purposes of the Union was to provide for the common defense," said Wigner. "It seems difficult to think of defense without making every effort toward protecting what is most important: the lives of the people."

Confrontations with the Soviet Union over Berlin and Cuba had spurred major funding for civil defense in the United States. Wigner met with the director of the Office of Civil Defense to propose use of the laboratory's talents in ecology, shielding, and radiation detection for civil defense research, and he spent the summer of 1963 leading a Defense Department seminar on civil defense problems.

With funding from the Office of Civil Defense assured, Wigner returned from Princeton University to ORNL in September 1964 for his third extended stay. He headed a staff of twenty, who operated on the premise that improved civil defense might reduce rather than increase the probability of nuclear war. Although outsiders disagreed, Wigner's group contended that civil defense could bolster disarmament negotiations because nations that had adequate civil defense could blunt the force of imprudent adventures.

The laboratory's civil defense research initially focused on underground tunnels to protect urban populations and on related issues of how to rid the tunnels of body heat, protect the passages from fire storms and blasts, and provide them with power, air, and other utilities. The researchers devised some ingenious solutions, such as storing blocks of ice underground to absorb body heat and supply water. From this base, their research expanded to include underground highways, subway systems, and parking garages as part of a protective system.

Designing such systems required demographic knowledge, such as the number and probable age distribution of the people to be protected. To uncover this information, the laboratory hired demographers Everett Lee and William Pendleton and joined Oak Ridge Associated Universities in sponsoring the formation of the Southern Regional Demographic Group in 1970.

The research also required an understanding of the reactions of people under the stresses that would accompany emergency use of underground shelters. To explore this problem, the laboratory hired its first social scientists, including Claire Nader, the sister of Ralph Nader.

Years before gaining fame as a consumer advocate, Ralph Nader came to the laboratory to write about its activities. Noting that Oak Ridge had not then attracted many technology firms such as those clustered near Boston and San Francisco, Nader asked whether its rural isolation was the culprit. "What the city-based people call our isolation, we call our freedom," responded one miffed Oak Ridge physicist, ". . . freedom from the congestion and implosion of the metropolis and freedom to match these beautiful natural environments between the Cumberlands and the Smokies with the finest possible work of our minds and hands."

The potential effects of nuclear fallout on this natural environment became a major concern of Stanley Auerbach and his radioecology scientists. Auerbach had attended early civil defense conferences with Wigner because of public concerns about the ecological consequences of a nuclear war. As one result, in 1967, small plots of land at the laboratory were treated with cesium-137–coated particles of weapons fallout size in order to observe their environmental effects. This proved to be the last large-scale field application of radionuclides at the laboratory, although radiotracer studies continued on previously contaminated sites. The bombs, of course, never came, but the ground on which the cesium was planted remains contaminated to this day—a reminder of the deadly fallout of radioactive substances and the domestic costs of nuclear research conducted in the 1950s and 1960s in the name of the Cold War.

After a year of setting the foundation for civil defense research, Eugene Wigner passed again through his revolving door back to Princeton University, promising to return regularly for additional defense consultations. James Bresee, and later Conrad Chester, succeeded Wigner as chief of the laboratory's civil defense research and, with added funding from the Department of Housing and Urban Development, explored multipurpose utility service tunnels, nuclear energy centers for cities, management of urban wastes, and a variety of other municipal problems related to emergency preparedness in the event of a nuclear attack.

Among the accomplishments of the civil defense staff were Joanne Gailar's analysis in 1969 of Soviet civil defense plans, which encouraged civilian evacuation planning in the United States to counter Soviet planning, and Cresson Kearny's field manual for survival skills and suitable shelters. Because underground shelters were energy-efficient, Kearny's manual subsequently enjoyed wide distribution. Studies examining the preservation of emergency food supplies and the needs for alternative energy sources, in fact, eventually brought the civil defense group an assignment to analyze solar energy.

During the late 1960s, Weinberg explored with the University of Tennessee and state officials the possibility of forming a Civil Defense Institute in Oak Ridge, similar to the Space Science Institute established at Tullahoma, Tennessee. This effort proved fruitless, but the laboratory's studies of emergency technology continued, concentrating on evacuation and sheltering from chemical hazards. At the outbreak of the 1991 Persian Gulf War, in fact, military authorities thought it worthwhile to dust off the laboratory's old civil defense reports on biological and chemical weapons.

Lab in Space

In 1961, Alvin Weinberg voiced concerns about a scientific Olympics with the Soviets that focused on launching manned spacecraft. He thought the space race had little connection with the well-being of people, and he worried about shielding spacecraft crews against solar radiation. NASA responded not by heeding Weinberg's worries but by funding ORNL studies of radiation shielding and the biological effects of solar radiation. NASA also partially funded the AEC Systems for Nuclear Auxiliary Power for long-distance space exploration.

The space race brought three million dollars into the ORNL budget in 1962, and by 1966, the laboratory had 160 personnel in ten different divisions participating in the space Olympics. Despite Weinberg's reservations, "space" soon emerged as a down-to-earth item on the laboratory budget sheets and personnel rosters, thus transcending more abstract policy concerns.

The Biology, Health Physics, and Neutron Physics divisions, for example, received assignments to assay the biological effects of radiation from the Van Allen belt and solar flares and to devise lightweight shields to protect the crews of the *Apollo* spacecraft. In addition to ground research, the Biology Division sent boxes containing bacteria and radioactive phosphorus aboard *Gemini 3* and *Gemini 11* and also placed blood samples aboard satellites to assess radiobiological effects in space. The Health Physics Division exposed small animals and plastic phantoms resembling humans to fast-burst radiation, thereby estimating the radiation dosages to internal organs that might await the *Apollo* crews. Fred Maienschein, Charles Clifford, and others in the Neutron Physics Division used data from the tower shielding facility and linear accelerators to design lightweight shielding for the *Apollo* spacecraft.

The AEC Systems for Nuclear Auxiliary Power program, begun in 1956, aimed to design compact, maintenance-free power generators for use in remote locations at sea, on land, and in space. Under AEC assignment, the laboratory undertook studies of two types of generators: miniature nuclear reactors and radioisotope generators.

Arthur Fraas led a team studying a small reactor using molten potassium to spin a turbine that could generate electricity for use in airless, weightless environments. Although not adopted by AEC for space missions, the reactor's boiling potassium technology found other applications.

The Isotopes Division received a major assignment from AEC to produce massive blocks and pellets of radioactive curium isotopes, which became incandescently hot as they decayed and provided power for thermoelectric generators. Most of these isotopes went into portable power generators built by the Martin Marietta Corporation to supply power to weather stations in the Arctic and to navy navigation buoys and beacons at sea. Because deep-space exploration required too many panels for the use of solar energy in the spacecraft, some tiny space

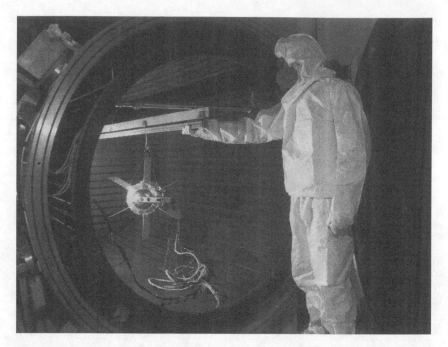

The laboratory during the 1960s and later assisted development of power sources for space probes.

probes launched toward the outer planets of the solar system during the 1970s used the Peltier effect from radioisotopic heat to produce electricity for as long as thirty years without fuel replenishment. These survey craft returned spectacular pictures of the outer planets back to Earth a decade or more later.

As planning for NASA missions to the moon began, the laboratory lost personnel to the space agency, including P. R. Bell, who, as director of NASA's Lunar Receiving Laboratory in Houston, requested assistance from his friends in Oak Ridge. Neil Armstrong in July 1969 and other astronauts who later landed on the moon carried telescoping scoops for collecting moon rocks designed by Union Carbide's General Engineering Division and fabricated by the Plant and Equipment Division in Oak Ridge. Richard Fox of the laboratory's Instrumentation and Controls Division—one of the veterans of the 1942 Fermi experiments in Chicago—designed the vacuum-sealed boxes that housed lunar rock samples after their return to Earth; some of those samples came to the laboratory for intensive study.

Although less than 4 percent of the laboratory's budget came from NASA programs, the personnel involved took pride in helping win the space race. In reflecting on the laboratory's work for NASA at the end of the 1960s, Weinberg ultimately observed that its scientific aspects had been challenging and its management even more so. NASA and other non-AEC projects, however, were subject to micromanagement by the agencies providing the funding. And the laboratory missed the budgetary flexibility that AEC-funded programs allowed.

The Lab and the Environment

Because AEC had no firm policy on performing work for other agencies, the laboratory during the 1960s approached each external effort ad hoc, gaining approval from AEC headquarters for each venture. By 1969, 14 percent of the laboratory's programs consisted of nonnuclear work for agencies other than AEC; at the time, Argonne, Brookhaven, and other laboratories had less than 1 percent of their work funded outside AEC.

In 1967, Congress amended the Atomic Energy Act to further encourage AEC laboratories to seek work from other government agencies. AEC, for instance, urged the laboratories to initiate studies of environmental pollution, then an increasingly favored and well-funded program under the Federal Water Pollution Control Agency. Weinberg advised AEC's general manager that Auerbach's ecological studies and Kraus's water research placed ORNL in a strategic position to attack water pollution by identifying pollutants and assessing their effects on aquatic and terrestrial life. Technology developed during the desalination studies, Weinberg and other laboratory officials contended, could be adapted to improve sewage wastewater treatment. Moreover, ORNL capabilities in analytical chemistry could be applied to investigations of atmospheric pollution, while biologists could expand their analysis of the effects of chemical agents on living organisms.

The Federal Water Pollution Control Agency did not accept the laboratory's first proposal in 1967 to investigate stream eutrophication and its relationship to farm land management. Auerbach and his ecologists then proposed to AEC that it approve ORNL investigations of the effects of heated water released from power plant cooling facilities into aquatic

Stanley Auerbach operates a hoist to move a cask of cesium-137-tagged sand during a 1967 ecological study.

systems. When AEC approved this initiative, Auerbach recruited Charles Coutant, an expert on aquatic thermal effects, to lead this research effort.

For environmental research at the laboratory, 1967 was literally and figuratively a watershed year. That year, AEC approved Daniel Nelson and James Curlin's proposed development of the Walker Branch Watershed research facility, a small stream basin near the main laboratory complex, as an experimental center for studies of relations between terrestrial and aquatic ecosystems.

With instruments above and below ground for precise measurements of stream flows, the Walker Branch facility, Auerbach later recalled, marked the beginning of educating ORNL personnel about the requirements of large-scale environmental research for sophisticated devices and instrumentation. In 1967 as well, the National Science Foundation (NSF) appointed Auerbach director of the ecosystems component of an International Biome Program for the eastern United States.

Funded at about one million dollars annually for eight years, this program was the first major funding by NSF of work at an AEC laboratory.

As the 1960s waned, national concerns about ecological damages and pollution threats made themselves felt in the political arena. As the environmental movement fermented, the laboratory's potential as a center for research relating to ecological problems received increasing recognition in official scientific circles. Auerbach, biologist William Russell, and other ORNL environmental and life scientists went on the road to public hearings, where they found people jittery about the environmental and health effects of nuclear energy. Although spearheading investigations of environmental pollution, the laboratory, along with AEC and the nuclear industry, found itself more and more on the defensive against charges leveled initially by environmental activists and subsequently by ordinary citizens. Questions concerning the safety of nuclear reactors thus became increasingly pertinent to ORNL research programs. What had been a popular struggle against Soviet science in the 1950s increasingly became a contentious campaign for credibility at home by the 1970s.

Nuclear Safety

In an effort to quell rising public distrust, by the end of the 1960s, 20 percent of the laboratory's reactor budget was devoted to nuclear safety. The laboratory operated a pilot plant to test fission product release and fuel transport and developed a heat-transfer facility to test fuel element behavior in the event of coolant-loss accidents. It also devised filters to contain radioactive iodine that might be released during accidents and participated in the design of auxiliary cooling systems for reactors to prevent meltdowns.

The laboratory's Heavy Section Steel Program, under Joel Witt and Graydon Whitman, closely examined reactor pressure vessels to ascertain their performance under various stresses. Early steel pressure vessels in reactors had ranged from three to ten inches thick, but the larger vessels designed by 1968 were as much as fourteen inches thick. The Heavy Section Steel Program's task was to investigate this armorlike steel and devise safety codes and standards for its use in reactor vessels.

Private nuclear industry shared the costs of heavy section steel investigations and other nuclear safety programs with AEC, but these studies were not considered work for other agencies. Instead, they were viewed as key ORNL initiatives, mandated by broad AEC responsibilities in nuclear power and safety.

To address possible future roles, the laboratory obtained NSF funding for summer seminars during the late 1960s. These seminars began in 1967 with a multidisciplinary study of a nuclear agro-industrial complex and expanded in 1968 to include ORNL, TVA, and university scientists and engineers investigating the resources of the Middle East and the health and education of its people. Milton Edlund and James Lane headed the Middle East studies for the laboratory and visited this distant region to explore potential developments there.

In the summers of 1969 and 1970, seminars organized by David Rose, who came to the laboratory from MIT, and by laboratory staff members John Gibbons, Claire Nader, and James Liverman addressed environmental issues and the general role of science in the formation of public policy. In retrospect, these far-ranging seminars were pivotal events in the formation of the laboratory's Environmental Sciences Division and Energy Division, which today employ many of the laboratory's social scientists. Out of these seminars, too, evolved a proposal to create national environmental laboratories, or at least one in Oak Ridge.

Declaring that "ecologists have displaced the physicists and the economists as high priests in this new era of environmental concern," Weinberg formed a National Environmental Concept Committee under David Rose. Rose wrote a report entitled *The Case for National Environmental Laboratories* and delivered a copy to Senator Howard Baker of Tennessee, who had it printed as a congressional document. Weinberg and Rose then met with Senators Baker and Edmund Muskie to discuss it. In early 1970, a House committee added four million dollars to the NSF budget earmarked for ORNL studies of sewage hyperfiltration, air pollution, waste management, and chemical toxicity, and Senators Baker and Muskie sponsored a resolution establishing a National Environmental Laboratory at Oak Ridge. Momentarily, it appeared that the laboratory might jump into the forefront of environmental science.

Representative Chet Holifield, a staunch ally of the laboratory and a key member of the Joint Committee on Atomic Energy, surprised the laboratory's staff when he blasted the Baker-Muskie resolution. Rumor had it that he said, "Let Muskie get his own laboratories!" Holifield added a rider to the 1970 AEC authorization that read:

> The Joint Committee sees signs that ambition to acquire new knowledge and expertise in fields outside the present competence and mission of an AEC National Laboratory, in order to attain and provide wisdom which this country needs in connection with non-nuclear environmental and ecological problems, is spurring at least one laboratory to solicit activities unrelated to its atomic energy programs and for which it does not now have special competence or talents.

Thus chastised by a powerful and influential legislator, Oak Ridge saw its chances of becoming the National Environmental Laboratory fatally wounded. Nevertheless, with enactment of the National Environmental Policy Act of 1970 and formation of the Environmental Protection Agency, the laboratory moved on a broader scale into environmental research. In March 1970, shortly before the first Earth Day celebrations, Weinberg expanded Auerbach's Ecology Section into an Ecological Sciences Division that embraced studies of terrestrial, aquatic, and forest ecology. At the same time, an environmental studies program funded by NSF and headed by John Gibbons applied social and economic expertise to energy-related environmental challenges.

With the addition of radiological assessment and geosciences groups, the Ecology Sciences Division became the Environmental Sciences Division in 1972. The national requirement that environmental impact statements be prepared for new federal projects brought the new division considerable work, and the division formed an Environmental Sciences Information Center to support the preparation of impact statements. It also participated in a multidisciplinary study, led by William Fulkerson, Wilbur Shults, and Robert Van Hook, of the environmental effects associated with fossil-fuel power plants.

In buildings constructed at the west end of the laboratory grounds, the expansion of environmental sciences at ORNL continued into the 1990s. In fact, if not in name, ORNL became a national environmental assessment laboratory.

Constraints

As early as 1967, Weinberg recognized that the costly Vietnam War was constraining the national budget for science. "Because of Vietnam, we shall be lucky to get as much money as we had this year," he told the staff. "We can only hope that Vietnam will be resolved quickly; and that, as peace is restored, we can devote ourselves and our expanding technologies to the creation of a better world."

The war did not end quickly and, in 1968, budgetary constraints forced retrenchments. Weinberg adamantly denied that the laboratory's nonnuclear efforts were intended to counter the reductions in nuclear science budgets; in fact, he reminded critics that those efforts had begun long before the budgetary shortfalls of the late 1960s. Although ORNL funding remained constant from 1965 to 1970, inflation eroded the funding's value by as much as 25 percent.

Other factors, in addition to the costs of the war, had a role in the declining budget. Because AEC was determined to proceed with the liquid-metal fast breeder reactor, it slashed funding from the laboratory's molten-salt thermal breeder program. As part of the social upheaval of the 1960s, strong antiscientific sentiment (fueled by a scientific establishment often oblivious—or insensitive—to public concerns) sparked rowdy confrontations, even at professional scientific conventions, which also affected congressional support for research to some extent.

Weinberg and laboratory staff saw several demonstrations against science by disillusioned youth. After seeing one in Boston in 1969, Weinberg wrote:

> We in Oak Ridge, living as we do in a sheltered and pleasant scientific lotus-land, just don't know what our colleagues in the beleaguered universities are up against. What a shock it is to go to the hub of the intellectual universe for what one expects to be a rather routine scientific meeting, and to run smack into a full-scale confrontation between the scientific establishment and the angry young people. I haven't had such an exciting time in years, certainly never at a scientific meeting.

At Christmas 1969, the federal government's Bureau of the Budget (today's Office of Management and Budget) ordered across-the-board cuts at the laboratory, reducing staff from 5,300 to less than 5,000. The laboratory's thermal breeder program was cut by two-thirds, and its proposed new particle accelerator, known as APACHE, was scrapped entirely. Departing friends made the 1969 holiday season in Oak Ridge as gloomy as that of 1947. In the close-knit Oak Ridge community, when friends lost their jobs, they usually had to leave to find work elsewhere.

"Our vast scientific apparatus is deployed against scientific problems—yet what bedevils us are strongly social problems," Weinberg lamented. "Can we somehow deploy our scientific instrumentalities, or invent new instrumentalities, that can make contributions to resolving these social questions?"

"We lost our innocence" about 1969, William Fulkerson, the laboratory's associate director for Advanced Energy Systems, recalled years later. Realizing that scientific problems had social and political contexts as well as technical components, the chastened laboratory entered the 1970s less innocent but more ready to meet the challenges of this tumultuous decade—one in which the nation would experience two energy crises and federally sponsored environmental programs and regulations that would forever alter the way the laboratory conducted its business.

Chapter 6

Responding to Social Needs

After thirty years of steady progress, AEC experienced an unexpected series of public events and controversies during the 1970s that dislodged society's confidence in nuclear energy. Events that affected the nuclear energy industry in general rippled through AEC, building to a tidal wave of discontent that ultimately led to dramatic changes in leadership within the agency.

Chaired from the early 1960s until the early 1970s by Glenn Seaborg, a Nobel laureate chemist associated with the Metallurgical Laboratory during World War II, AEC was led subsequently by an economist, and then a marine biologist, before being split into the Energy Research and Development Administration (ERDA) and the Nuclear Regulatory Commission (NRC) in 1974. This division confirmed that the institutional framework, which had served nuclear power well in the years following World War II, would be insufficient to meet the energy challenges of the future.

If events within AEC mirrored larger trends within the nuclear energy industry, then it also can be said that the laboratory reacted to the dramatic transitions within AEC with its own series of critical changes. Although not sundered like AEC, the laboratory transcended its traditional focus on uranium fission to undertake broader missions that encompassed all forms of energy. At the same time, ORNL leadership passed from the hands of a fission expert to a nuclear fuel-reprocessing specialist and, finally, to an expert in fusion energy.

With more powerful research reactors and accelerators added to its fleet during the 1960s, the laboratory became a premier international center for producing and separating transuranic elements. Researchers studied the structures of transuranic elements and nuclei using accelerated particles ranging in mass from protons through curium ions. In support of the AEC reactor program, the laboratory specifically pursued development of a molten-salt reactor and generally continued to investigate liquid-metal and gas-cooled reactor technologies. By 1970, in response to the new political realities that the nuclear industry faced, the laboratory also became a center for exploring the safety, environmental, and waste-disposal challenges presented by nuclear energy.

The laboratory's advance into new research frontiers was both a response to necessity and a deliberate effort to assume new challenges. Budget shortfalls between 1969 and 1973 shelved ORNL plans for new reactors and reduced its staff from nearly 5,500 in 1968 to fewer than 3,800 by 1973. Moreover, the laboratory's wartime veterans, now in their fifties and sixties, began to retire as the laboratory's thirtieth anniversary neared in 1973.

The departure of Oak Ridge's Manhattan Project engineers and scientists left a void in the laboratory's institutional culture that was filled by members of a new generation who brought their own interests and experiences to the research agenda. Having come of age in the 1960s, this new generation carried somewhat different priorities and sensibilities to the workplace than had the laboratory's original scientists, for whom World War II served as the defining moment in their careers.

To meet these challenges, laboratory management reorganized and launched a series of retraining programs designed to transcend the laboratory's traditional uranium fission focus. These new efforts led to investigations into all forms of energy—a broadening of research that sought to make the laboratory responsive to the political and social changes sweeping the nation.

In the aftermath of Earth Day in April 1970 and the passage of a series of environmental laws and regulations intended to bring environmental concerns to the forefront of national policy, the public clamored for more "socially relevant" science that would address everyday concerns. In 1973, as Americans lined up to purchase gasoline and turned down their thermostats to compensate for heating oil shortages, the desire for relevant science was never more urgent.

The high-flux isotope reactor and adjacent facilities for processing transuranium elements in Oak Ridge.

ORNL efforts to explore new, nonnuclear energy issues during the early 1970s proved to be both timely and critical. Born at the dawn of the nuclear age and nurtured to maturity during nuclear power's great leap forward in the 1950s, the laboratory was not about to abandon its ties to nuclear research. Nevertheless, as it experienced and then responded to the dramatic changes of the 1970s, it emerged from this tumultuous decade a multipurpose science research facility, ready to tackle the increasingly complex issues of energy and the environment.

Super-Duper Cooker

The high-flux reactor designed under Eugene Wigner's supervision in 1947 and built in Idaho provided the highest neutron flux then available. By the late 1950s, however, the Soviets had designed a reactor that surpassed it.

"We do not believe the United States can long endure the situation of not having the very best irradiation facilities in the world at its disposal," commented Clark Center, Union Carbide chief at Oak Ridge. "Therefore, we would like to suggest that the Atomic Energy Commission undertake actively a design and development program aimed at the early construction of a very high-flux research reactor." Glenn Seaborg, an expert in transuranic chemistry, concurred with Center and urged AEC to build a higher-flux reactor.

With these statements of support echoing in Washington, Weinberg brought Wigner back to ORNL to discuss the design of a more powerful reactor, which Weinberg labeled a "super-duper cooker." Trapping a reactor neutron flux inside a cylinder encasing water-cooled targets, this high-flux isotope reactor would make possible "purely scientific studies of the transuranic elements" and augment the "production of . . . radioisotopes." Weinberg also insisted that the reactor be built with beam ports to provide access for experiments.

Charles Winters, Alfred Boch, Thomas Cole, Richard Cheverton, and George Adamson led the design, engineering, and metallurgical teams for this hundred-megawatt reactor, completed in 1965 as the centerpiece of the laboratory's new transuranium facilities. Seaborg, appointed AEC chairman by President Kennedy, returned to ORNL in November 1966 for the dedication. He declared that the exotic experiments made possible by this new high-flux isotope reactor would "deepen our comprehension of nature by increasing our understanding of atomic and nuclear structure."

Built in Melton Valley, across a ridge from the main X-10 site in Bethel Valley, the high-flux isotope reactor irradiated targets to produce elements heavier than uranium at the upper and open end of the periodic table. At a heavily shielded transuranium processing plant adjacent to the reactor, A. L. Lotts of the Metals and Ceramics Division led the teams fabricating targets that would subsequently be inserted into the reactor.

In the reactor, the targets were placed in the high neutron flux, where they absorbed several neutrons in succession, making their way up the periodic table as they increased in mass and charge. Then the irradiated targets were returned to the processing plant for chemical extraction of the heavy elements berkelium, californium, einsteinium, and fermium.

Previously available only in microscopic quantities, the milligrams of heavy elements produced at the high-flux isotope reactor proved immensely valuable for research. The laboratory, in fact, distributed the heavy elements to scientists throughout the world and to its own scientists housed in the new transuranium research laboratory. "Our main effort at ORNL," said Lewin Keller, head of transuranium research, "is directed toward ferreting out their nuclear and chemical properties in order to lay a base for a general understanding of the field."

Of the transuranic elements, an isotope of element 98 garnered greatest attention. Named for the state where it was discovered, californium-252 fissions spontaneously to provide an intense neutron source able to penetrate thick containers and induce fission in uranium-235 and plutonium-239. The process produces short-lived, on-site isotopes in hospitals for immediate use in patients. Cancerous tumors could now be treated by implanting californium needles instead of less effective radium needles that were used previously. Other transuranic elements afforded practical applications such as tracers for oil well exploration and mineral prospecting.

Thanks to Weinberg's foresight in demanding beam ports, the high-flux isotope reactor could be used for important investigations of materials by neutron-scattering techniques. Studies were made of the magnetic properties and crystal structures of various materials by Wallace Koehler, Michael Wilkinson, Henri Levy, and their associates in the Solid State and Chemistry divisions. The intense neutron beams from the reactor coupled with state-of-the-art neutron-scattering instrumentation allowed never-before experiments to be performed. Of particular significance were materials investigations focusing on the magnetic interactions of neutrons with materials, which helped explain some unusual magnetic properties of rare-earth metals, alloys, and compounds.

The high-flux isotope reactor served science, industry, and medicine well for a quarter century. Although shut down because of vessel embrittlement in November 1986 and subsequently restarted at 85 percent of its original power, by 1991 it had gone through three hundred fuel cycles, generating benefits that ranged from advancing knowledge of materials by neutron scattering to enhancing understanding of U.S. history.

In 1991, for example, the high-flux isotope reactor's neutrons supported activation analysis of hair and nail samples from the grave of

Larry Robinson and Frank Dyer examine a sample of hair removed from the body of President Zachary Taylor.

President Zachary Taylor, which indicated he had not been poisoned by arsenic while in office, as one historian suspected. Americans could rest assured that President Taylor had died of natural causes and not a conspiracy—thanks to the use of late-twentieth-century technology in the service of early-nineteenth-century history.

The Last Reactors

Between the 1940s and 1960s, the construction of new reactors was part of the laboratory's ever-changing landscape. The reactors built in the 1960s, however, would mark the end of ORNL's "bricks and mortar" reactor era. No new reactors would be built during the 1970s and 1980s, a remarkable dry spell given the rapidly changing nature of nuclear research.

Before being forced to close its doors on new reactor construction, the laboratory (in addition to its work on the high-flux isotope reactor)

AEC's experimental gas-cooled reactor was built during the 1960s in Oak Ridge.

completed the health physics research reactor, experimented with a molten-salt reactor, and performed research for AEC programs to develop a liquid-metal fast breeder reactor and for high-temperature gas-cooled reactors that stirred the interest of the private sector. Next to the high-flux isotope reactor, the most successful laboratory reactor built during this decade was the health physics research reactor.

Known originally as the "fast burst reactor," the health physics research reactor was installed in the new dosimetry applications research facility in 1962. John Auxier, later director of the Health Physics Division, managed the design and operation of this small, unmoderated, and unshielded reactor.

Composed of a uranium-molybdenum alloy and placed in a cylinder eight inches high and eight inches in diameter, the reactor required the insertion of a rod into the cylinder to release a neutron pulse used for health physics and biochemical research. In particular, data from research using the reactor, which remained operational until 1987, pro-

Craftsmen in 1975 adjust a testing machine for the liquid-metal fast breeder reactor program at ORNL.

vided guidance for radiation instrument development and dosage assessment. During the 1960s, for example, it helped scientists estimate the solar radiation doses to which *Apollo* astronauts would be subject.

When AEC suspended work on the experimental gas-cooled reactor in 1964, light-water reactors became the dominant sources of commercial nuclear power. As a result, the laboratory's gas-cooled reactor research waned. When Gulf General Atomic Corporation obtained orders for four high-temperature gas-cooled reactors in 1972, however, AEC boosted ORNL research funds for this technology. Specifically, the laboratory tested graphite-coated particles to fuel gas-cooled reactors. In addition, ORNL studies of thorium fuel recycling accelerated because gas-cooled reactors could use uranium-233 derived from thorium as fuel.

Laboratory research into the liquid-metal fast breeder reactor, which had been developed at Argonne National Laboratory, expanded during the late 1960s. William Harms coordinated ORNL's breeder technology program. His staff simulated the fast breeder's fuel assemblies,

AEC Chairman Glenn Seaborg starts the molten-salt reactor at ORNL in 1968.

using electric heaters, and tested reactor coolant flows and temperatures. A metallurgical team headed by Peter Patriarca evaluated the materials to be used in the fast breeder's heat exchangers and steam generator.

Work on the breeder accelerated in 1972, when AEC made Oak
Ridge the site of AEC's demonstration fast breeder reactor. Laboratory
efforts continued until Congress canceled the project in the mid-1980s,
after more than a decade of political controversy and debate amid the
gradual realization that the United States would not need a breeder reactor for twenty or more years because of the low cost and ready availability
of uranium. With ample supplies of nuclear fuel, the breeder's potential to
accelerate the proliferation of nuclear weapons (by creating more nuclear
material than it consumed) hardly seemed worth the risk.

"Dark Horse" Breeder

"A dark horse in the reactor sweepstakes"—that's how Alvin Weinberg
once described the laboratory's molten-salt reactor experiment to Glenn

Seaborg. Weinberg explained that if Argonne's fast breeder encountered unexpected scientific difficulties, Oak Ridge's molten-salt thermal breeder could serve as a backup that would help keep AEC's research efforts on track.

Based on technology developed for the aircraft nuclear reactor, molten-salt reactor experiments were conducted in the same building that had housed the aircraft reactor. Following the design and construction phases, molten-salt reactor experiments began in June 1965. Project directors Herbert MacPherson, Beecher Briggs, and Murray Rosenthal successively supervised experiments using uranium-235 fuel.

When the fuel was changed to uranium-233 in October 1968, AEC Chairman Seaborg joined Raymond Stoughton, the ORNL chemist who codiscovered uranium-233, to raise the reactor to full power. "From here," said Rosenthal, "we hope to go on to the construction of a breeder reactor experiment that we believe can be a stepping stone to an almost inexhaustible source of low-cost energy."

Weinberg and the laboratory's staff pressed AEC for approval of a molten-salt breeder pilot plant. They hoped to set up the pilot plant in the same building that had housed AEC's experimental gas-cooled reactor until that project was suspended in 1966.

Argonne's fast breeder had the momentum, however, and Congress proved unreceptive to laboratory requests to fund large-scale development of a molten-salt breeder. Appealing personally to Seaborg, a chemist, Weinberg complained, "Our problem is not that our idea is a poor one—rather it is different from the main line, and has too chemical a flavor to be fully appreciated by nonchemists."

Meanwhile, the experimental molten-salt reactor operated successfully on uranium-233 fuel from October 1968 until December 1969, when the laboratory exhausted project funds and placed the reactor on standby. The laboratory continued molten-salt reactor research, as limited funding allowed, until January 1973, when the AEC reactor division abruptly ordered work to end within three weeks.

In the wake of the energy crisis in late 1973, however, new funding for molten-salt research emerged and continued until 1976. An exclusive ORNL project, the molten-salt reactor, in Weinberg's opinion, was the laboratory's greatest technical achievement. His reasoning was based on the following observations: The molten-salt reactor was feasible, could use uranium-233 made from abundant thorium as fuel, and offered greater safety than most other reactor types.

As late as 1977, an electric utility executive advised President Carter of his company's interest in a commercial demonstration of the molten-salt breeder reactor. The government's preoccupation with the ill-fated liquid-metal fast breeder reactor, however, drove Oak Ridge's thermal breeder into obscurity. To Weinberg's chagrin, the "dark horse" reactor never emerged from the pack to lead the nuclear research effort.

Accelerators

An evolution similar to the molten-salt breeder program marked the laboratory's accelerator program of the 1960s. The laboratory's advanced particle accelerators, an isochronous cyclotron and an electron linear accelerator, moved the ORNL accelerator program to the forefront of the nation's research efforts in this field. However, competition from other accelerator projects as well as funding constraints would stall the program in the early 1970s.

The Oak Ridge Isochronous Cyclotron (ORIC) began operating in 1963, firing protons, alpha particles, and other light projectiles into targets to produce heavy ions. To compensate for increases in the mass of ions as they are accelerated, the cyclotron had an azimuthally varying, but radially increasing, magnetic field to focus the particle's paths and keep them in resonance at high energies. In its day, ORIC was first of a kind and a major technological breakthrough.

Built on the eastern side of the X-10 site in Bethel Valley, the new cyclotron brought Robert Livingston's Electronuclear Division from the Y-12 to the X-10 site. In 1972, the Electronuclear Division consolidated with the Physics Division under the direction first of Joseph Fowler, followed by Paul Stelson and James Ball, reporting to Alex Zucker, the associate director for physical sciences.

A year after ORIC obtained its first heavy-ion beam, the laboratory completed its Oak Ridge Electron Linear Accelerator (ORELA). Except for an office and laboratory building, this accelerator was underground, covered by twenty feet of earth shielding. Electron bursts traveled seventy-five feet along the accelerator tube to bombard a water-cooled tantalum target, producing ten times as many neutrons for short pulse operation than any other linear accelerator in the world. From the target room, the neutrons passed through eleven radial flight tubes to underground stations for experiments.

A joint project of the Physics and Neutron Physics divisions, ORELA's main purpose was to obtain fast-neutron cross-sections for the fast-breeder reactor program. It served this purpose admirably, contributing a great deal to fundamental physical science in the process. In 1990, for example, ORELA's intense neutron beams bombarded a lead-208 target and measured the separation between the three quarks composing a neutron. This research effort advanced scientific understanding of the strong force that glues a neutron together.

By the time ORIC and ORELA were fully operational in 1969, the laboratory had planned to build another machine capable of accelerating heavy ions into an energy range where superheavy transuranic elements could be investigated. With the support of universities throughout the region, this accelerator began as a southern regional project. In fact, the laboratory considered naming it CHEROKEE (after one of the Southeast's most noted Native American tribes), but top scientists could not find the words to form an appropriate acronym; so they named it APACHE, the Accelerator for Physics and Chemistry of Heavy Elements.

Blanching at its twenty-five-million-dollar cost, President Richard Nixon's budget officers rejected the laboratory's regional APACHE concept in 1969. Discussing the administration's unfavorable decision at AEC headquarters, Alex Zucker learned the budget office and AEC would consider only national, not regionally sponsored, accelerators. To secure approval for an advanced accelerator, it would be necessary for the laboratory to explain the unmet challenges of heavy-ion research, show that it served "truly important national needs," and demonstrate that it would protect the United States from being surpassed in scientific research by other nations, particularly the Soviet Union.

Asserting that the proposed accelerator would advance understanding of "the behavior of nuclei in close collision and the properties of highly excited, very heavy nuclear aggregates," Zucker recommended that the laboratory recast its new accelerator project in broader terms, naming it the National Heavy Ion Laboratory. Accepting this counsel, Weinberg established a steering committee headed by Paul Stelson to reformu-

The ELMO Bumpy Torus built at the laboratory for fusion experiments.

late the proposal. The committee's efforts were fostered by university physicists from the South who saw value in having the accelerator located in Oak Ridge.

Led by physicists Joseph Hamilton of Vanderbilt University and William Bugg of the University of Tennessee, a consortium formed in 1968 to unite physicists from eighteen universities interested in heavy-ion research at ORIC and the laboratory's proposed national accelerator. Working with Robert Livingston and Zucker, the consortium obtained combined funding from their universities, state government, and AEC to finance the construction of an addition to the ORIC building. The addition housed the University Isotope Separator (UNISOR), which interfaced with ORIC beam lines.

Only the Soviet Union had another on-line separator connected to a heavy-ion accelerator. Equally important, this effort represented the first combined funding project for nuclear research hardware in the United States. When the separator facility was completed in 1972, UNISOR's consortium scientists initiated investigations into new radioisotopes for ini-

The interior of ORMAK, the laboratory's first tokamak for fusion experiments.

tiatives as practical as cancer research and as abstract as the study of heavy nuclei generation in the stars.

UNISOR and ORIC's ongoing research and widespread academic participation gave the laboratory proof that its proposed National Heavy Ion Laboratory would serve national needs. Budgetary constraints, however, delayed approval of this new facility until 1974. Named the Holifield Heavy Ion Research Facility after Congressman Chet Holifield, the long-time chairman of the Joint Committee on Atomic Energy, this new accelerator was completed in 1980. It attracted scientists from throughout the world to Oak Ridge.

Gold-Plated Fusion

Although the laboratory's molten-salt breeder and APACHE accelerator hit fiscal walls in 1969, its fusion energy research continued to receive funding under the stimulus of international competition. In 1969,

AEC authorized the laboratory to construct a gold-plated fusion machine called ORMAK.

After a wildly optimistic, but essentially unsuccessful, entry into fusion energy research in the 1950s, the world's scientists recognized that better understanding of hydrogen plasma behavior was necessary before any real progress could be made. As a result, fusion scientists settled into the computer trenches during the 1960s, hoping to improve the theoretical underpinnings of fusion energy in an effort to provide a firmer foundation for laboratory studies and demonstrations. When it came to fusion, scientists faced a fundamental shortcoming that had to be overcome: They were, to put it simply, unsure of how to make it work in practical terms.

At the laboratory, attention focused on microinstabilities associated with the electric fields within the plasma of fusion devices. Empirical experiments continued with both a second direct current experiment and a steady-state fusion device conceived by Raymond Dandl and given the odd name ELMO Bumpy Torus. ELMO's electron cyclotron heating set a record for steady, stable hot-electron plasma but still fell far short of a performance record that even hinted of practical applications.

Optimism about fusion nevertheless resurfaced in 1968, when Soviet scientist L. A. Artsimovich of Moscow's Kurchatov Institute announced that his doughnut-shaped tokamak had confined a hot plasma. When Artsimovich visited the United States in 1969, Herman Postma, ORNL chief of fusion research, dispatched an ORNL team to discuss tokamaks with him.

Enthusiastic about what they heard, Postma's team proposed to AEC that the laboratory be given authorization to construct a tokamak. They received quick approval, with a mandate to have it operational by 1971. While the Oak Ridge tokamak, called ORMAK, brought the laboratory back into a race with the Soviets, Artsimovich and other Soviets, in the unique cooperative spirit that characterized fusion research even during the Cold War, provided helpful information for ORMAK's design.

Sometimes working three shifts daily, the laboratory's thermonuclear staff, with assistance from skilled craftspeople at Y-12, rushed ORMAK's construction. The plasma was created inside a doughnut-shaped vacuum chamber (torus) of aluminum with a gold-plated liner. Coils of electrical conductors cooled by liquid nitrogen created the magnetic field.

Michael Roberts, ORMAK's project leader, described the assembly of
this complicated machine as an unusual exercise like "putting an orange inside an orange inside an orange, all from the outside."

In the summer of 1971, ORMAK generated its first plasma and experiments began, with encouraging results achieved by 1973. Herman
Postma worried, however, whether the high-speed neutrons they generated would destroy the fusion reactors. Materials had to be found for
fusion reactor walls that would withstand the particle damage and
stresses before the ORMAK or other fusion devices could generate even
a shimmer of interest among commercial power producers.

More optimistic, Weinberg noted that the ORMAK design permitted installation of a larger torus that would become ORMAK II. "With
great good luck," he forecast, "ORMAK II might tell us that it would be a
good gamble to go to a big ORMAK III, which might be the fusion equivalent of the 1942 experiment at Stagg Field in Chicago."

Elusive plasma slipped from ORMAK's golden grip, however, and
neither ORMAK nor subsequent fusion machines have yet achieved a self-
sustaining fusion reaction. Experiments at Princeton University in the fall
of 1993 again spiked the optimism of fusion scientists and garnered worldwide attention. But the experiment could be sustained for only a few minutes and consumed twice as much energy as it generated.

Nuclear Energy and the Environment

While basic science and experimental reactor and accelerator hardware
dominated activities within the laboratory, political, legal, and citizen
protests far from the Oak Ridge reservation contributed mightily toward reorienting its missions after 1969. Although dozens of reactors for
commercial power production were then in the planning and construction phases, the nuclear industry remained troubled by three concerns: reactor safety, power plant environmental and public health impacts, and
safe disposal of radioactive wastes. These concerns also challenged the
laboratory.

After thirteen years of study, the laboratory proposed entombing
high-level radioactive wastes in deep salt mines near Lyons, Kansas. In
1970, AEC provided twenty-five million dollars to proceed with the salt

mine repository project. Noting that the wastes would be hazardous for thousands of years, Weinberg warned, "We must be as certain as one can possibly be of anything that the wastes, once sequestered in the salt, can under no conceivable circumstances come in contact with the biosphere." ORNL scientists concluded that the salt mines, located in a geologically stable region, would not be affected by earthquakes, migrating groundwater, or even continental ice sheets that might reappear during the waste's long-lived radioactivity.

People living near Lyons initially supported the laboratory's salt vault plan, but environmental activists and Kansas state officials opposed use of the salt mines on several grounds. Their concerns extended beyond questions of technical capability to deep-seated worries about sound and effective administration over the long haul. Activists claimed that underground disposal for millennia would require the creation of a secular, multigenerational "priesthood" charged with warning people never to drill or disturb the burial grounds—a level of vigilance difficult to sustain and perhaps undemocratic even if successful. "It is our belief that disposal in salt is essentially foolproof," replied Weinberg, although conceding that a "kind of minimal priesthood will be necessary."

During intense design studies in 1971, the laboratory and its consultants found that the many well holes already drilled into the Lyons salt formation in some circumstances might allow groundwater to enter the salt mines (such as during periods of excessive rain), thus raising technical questions about the site's long-term suitability. The salt mine disposal plan also became a heated political issue in Kansas, with leading state officials voicing objections to having their state targeted as the nation's high-level nuclear-waste dump. In 1972, AEC authorized the Kansas geological commission to search for alternative salt mines in Kansas and directed the laboratory to study salt formations in other states.

For the moment, AEC announced, high-level radioactive wastes would be solidified and stored in aboveground concrete vaults at the site of their origin. That moment has turned into decades, as scientific and political debates concerning radioactive waste-disposal issues continue to this day. With public distrust and animosities toward federal agencies responsible for overseeing nuclear-waste management running white-hot, the waste issue is not likely to be resolved soon.

During the early 1970s, public and legal concerns about the environmental effects of nuclear power brought the laboratory's studies of terrestrial and aquatic habitats to the forefront of its research agenda. Using the "systems ecology" paradigm pioneered by Jerry Olson, ORNL ecologists investigated radionuclide transport through the environment. In his pioneering work, Olson examined the migration of cesium-137 through forest ecosystems by inoculating tulip poplar trees behind the health physics research reactor with cesium-137, thereby establishing the first experimental research center for forest ecosystem studies.

In 1968, NSF placed Stanley Auerbach in charge of a deciduous forest biome program in which ORNL contracted with universities for studies of photosynthesis, transpiration, insects, soil decomposition, and nutrient cycling in forest systems in the eastern United States. That same year, David Reichle led the laboratory's forest research team that initiated large-scale forest ecosystem research. This was the forerunner of the laboratory's programs for investigation of acidic deposition, biomass energy production, and global climatic change.

Environmental studies at the laboratory received an unexpected boost in 1971 when a federal court, in a decision on a planned nuclear plant at Calvert Cliffs, Maryland, ordered major revisions of AEC environmental impact statements as an essential part of reactor licensing procedures. Required to complete almost one hundred environmental impact statements by 1972, AEC asked for help from its Battelle Northwest, Argonne, and Oak Ridge national laboratories. Giving this effort the highest priority, Weinberg declared, "Nuclear energy, in fact any energy, in the United States simply must come to some terms with the environment."

The laboratory's skeleton staff for environmental impact statements, headed by Edward Struxness and Thomas Row, expanded in 1972 to include seventy scientists and technicians, divided into discipline-oriented teams to rapidly prepare these applied ecology and socioeconomic reports. The staff who worked on these reports formed the nucleus of the Energy Division, established in 1974 under Samuel Beall's leadership.

The Calvert Cliffs decision required AEC to consider the effects of the heated water discharges from nuclear plants on the aquatic environment, and Charles Coutant led an ORNL team assigned the task of developing federal water temperature criteria to protect aquatic life.

Ecologist Carolyn Young in 1977 studies the effects of warm discharges from the Bull Run steam plant on milfoil in Melton Hill Lake.

For these and related studies, the laboratory initiated construction of an Aquatic Ecology Laboratory, completed in 1973.

The Aquatic Ecology Laboratory's initial equipment consisted of twenty water tanks, each containing various fish species, and a computer-controlled circulating heated water system to supply water of the proper temperature to the tanks; outside were six ponds for breeding fish and conducting field experiments. Initial researchers at the Aquatics Ecology Laboratory investigated the survival rate of fish and fish eggs at elevated temperatures. Only the Pacific Northwest Laboratory had a similar facility.

An indirect result of the aquatic studies came during licensing hearings for the Indian Point-2 nuclear plant on the Hudson River north of New York City. Because the Environmental Sciences Division identified Indian Point as a major spawning ground for striped bass, the impact statement for Indian Point-2 called for closed-cycle cooling towers to protect aquatic life from the adverse effects of thermal discharges. This decision, based on ecosystem modeling of the striped bass by Siegurd

Christensen and Webb Van Winkle, is considered one of the high points of environmental impact statement preparation at the laboratory.

The expense of environmental mitigation, reflected in the costs of constructing water cooling systems, concerned many nuclear power advocates. They were troubled as well by the stringent reactor safety standards that the laboratory staff proposed in 1970. Under the direction of Myer Bender of the General Engineering Division, the laboratory had recommended nearly one hundred interim safety standards. Many of these standards were based on investigations by the Heavy Section Steel Program conducted in the Reactor and Metals and Ceramics divisions. Other standards relating to reactor controls were developed by the Instrumentation and Controls Division.

William Unger and his associates, for example, designed and tested shipping containers for radioactive materials to determine the design that could best withstand collisions during transport. Richard Lyon and Graydon Whitman assessed the ability of reactors to withstand earthquakes, joining with soil engineers who simulated mini-earthquakes by detonating dynamite near the abandoned gas-cooled reactor. George Parker's team studied fission product releases from molten fuels, and Philip Rittenhouse's team investigated the potential failure of engineered safeguards, particularly the effects of interruptions in the water flow to reactors.

Emergency Core Cooling Hearings

"We find ourselves increasingly at those critical intersections of technology and society which underlie some of our country's primary social concerns," Weinberg declared in 1972. He also noted that laboratory veterans longed for the days when "what we did at ORNL was separate plutonium, measure cross-sections, and develop instruments for detecting radiation."

Those happy days, which were part of the laboratory's history, were now overshadowed in the heated climate of political discourse and public opinion that emerged during the Emergency Core Cooling Systems (ECCS) hearings in 1972. The AEC Hearings on Acceptance Criteria for Emergency Core Cooling Systems for Light-Water–Cooled

Nuclear Power Reactors, called the ECCS hearings for short, proved a critical event, one that forced the laboratory to face the harsh realities of the new nuclear era of controversy, conflict, caution, and compromise.

In 1971, President Nixon appointed James Schlesinger, an economist from his budget office, to succeed Glenn Seaborg as AEC chairman. Schlesinger aimed to convert AEC from an agency that unabashedly promoted nuclear power to one that served as an unbiased "referee." When protest greeted AEC's interim criteria for emergency core cooling systems, he convened a quasi-legal hearing for comments from reactor manufacturers, electric utility officials, nuclear scientists, environmentalists, and the public. The hearing began in Bethesda, Maryland, in January 1972 and lasted throughout the year.

To present their views, environmental groups hired attorneys and scientific consultants, who joined attorneys for reactor manufacturers, utilities, and the government to pack the ECCS hearings. Witnesses were subjected to dramatic cross-examinations—a new experience for most scientists, who were accustomed to establishing scientific "truth" through the sedate publications and peer review process, not through raucous adversarial legal proceedings.

As long as nuclear reactor power was less than four hundred thermal megawatts, a reactor's containment vessels could prevent a meltdown, or the type of accident popularly called the "China syndrome." Once reactors of greater power were designed, however, the containment vessel no longer could be counted on as the final defense, and an emergency core cooling system became imperative to protect the public. Weinberg thought it unfortunate that some AEC staff members had not been impressed by the seriousness of this requirement until forced to confront it by activists opposed to nuclear energy altogether.

Schlesinger agreed with Weinberg that ORNL staff should present their expertise fully and without reservation, regardless of whether they agreed with the interim criteria. Weinberg complained, however, that his staff should have been involved as fully in preparing the interim criteria as they would be in testifying at the hearings.

Among laboratory staff participating in these lengthy, sometimes contentious, sometimes tedious hearings were William Cottrell, Philip Rittenhouse, David Hobson, and George Lawson. They and other witnesses were grilled by attorneys for days. More than twenty thousand pages of testimony were taken from scientists and engineers, who often expressed sharp dissent on technical matters concerning the adequacy of the safety program. ORNL experts generally considered that existing criteria for reactor safety were based on inadequate research.

As a result of these showdown hearings, in 1973 AEC tightened its reactor safety requirements to reduce the chances that reactor cores would overheat as a result of loss of the cooling water. This measure, however, failed to placate critics who preferred a moratorium on nuclear reactor construction.

ORNL emphasis on reactor safety and environmental protection made the laboratory and Director Weinberg unpopular among some nuclear power advocates and members of the AEC staff—a strange turn of events for scientists who had devoted their careers to inventing and advancing practical applications of nuclear energy. Opponents of nuclear power, on the other hand, enjoyed quoting Weinberg's chilling declaration:

> Nuclear people have made a Faustian Contract with society; we offer an
> almost unique possibility for a technologically abundant world for the
> oncoming billions, through our miraculous, inexhaustible energy source;
> but this energy source at the same time is tainted with potential side
> effects that if uncontrolled, could spell disaster.

Although other events and considerations also played a part, the ECCS hearings of 1972 no doubt weighed heavily on the major management shifts in 1973 at the laboratory and AEC. They certainly influenced the decision by the president and Congress to divide AEC functions, separating its regulatory responsibilities from its other, more development-oriented activities. As a result, a profound transition in federally sponsored energy research and development began and continued throughout the mid-1970s.

Energy Transition

Another crisis—not in public confidence but in energy supplies—threatened the nation during the early 1970s. To meet this challenge, Weinberg sought to reorient and broaden the laboratory's mission. He was encouraged by both NSF and AEC, which in 1971 received congressional approval to investigate energy sources other than nuclear fission. At AEC headquarters, James Bresee, who had headed the laboratory's civil defense studies, became head of a general energy department, which managed funding for Oak Ridge's innovative energy studies.

When Congress authorized AEC in 1971 to investigate all energy sources, Weinberg appointed Sheldon Datz and Michael Wilkinson as the directors of a committee to review opportunities for nonnuclear energy research in the basic physical sciences. In addition, he made Robert Livingston the head of an energy council assigned the task of considering new laboratory missions.

At AEC, James Bresee reviewed ORNL proposals for wide-ranging research initiatives, including investigations into improved turbine efficiency, coal gasification, high-temperature batteries, and synthetic fuels to replace petroleum and natural gas.

The AEC-sponsored studies complemented related studies begun in 1971 under the auspices of NSF. Charged with sponsoring "socially relevant" science, NSF in 1970 sponsored interdisciplinary research at the laboratory slanted toward addressing broad societal problems. Led by David Rose and John Gibbons, this research focused in part on renewable resources, recycling, and regional modeling to determine future resource use. In broad terms, it sought to identify nationally significant energy and environmental problems that the laboratory could address.

When NSF announced its "research applied to national needs," or RANN, program in 1971, Weinberg advised NSF director William McElroy that the laboratory had "rather miraculously" identified many national needs for research that it could conduct. A poll of ORNL staff produced 150 new energy and environmental research proposals; NSF approved several of them. These efforts were guided by Wilbur Shults, Robert Van Hook, and William Fulkerson.

Noting that many environmental problems arose as a result of increasing energy use, Roger Carlsmith, Eric Hirst, and their associates, for example, initiated studies that examined ways to reduce energy demand by promoting energy conservation. They asserted that better home insulation could substantially cut energy use for home heating. Moreover, they concluded that increasing the efficiency of transportation and home appliances could significantly lower levels of energy consumption. These early efforts launched the laboratory's energy conservation research efforts, which would become one of its strengths in the postnuclear era.

To promote the design of more efficient central power stations, the laboratory studied improved turbine cycles, cryogenic power transmission lines, and "power parks" to cluster power stations outside urban areas. Arthur Fraas and his associates, for example, applied the potassium vapor technology they had developed for spacecraft to improve turbine efficiency at power stations.

Interest in solar energy flared in 1971, when solar energy advocate Aden Meinel visited the laboratory and proposed using solar energy to heat liquid-sodium and molten salts for large-scale generation of electricity. Murray Rosenthal, who headed the laboratory's molten-salt reactor experiment, led a group that assessed the economics of using heat from the sun to produce electricity.

Although the group concluded that solar power generation would cost more than nuclear or fossil fuel power, Rosenthal recommended additional studies because solar energy could ultimately prove economically attractive if two possible scenarios became a reality: "One is that environmental concerns or other factors could increase coal and nuclear energy costs more than we can foresee; the other is that the collection and conversion of solar energy could become much less costly than we assume."

With NSF backing, the laboratory examined solar energy as a potential long-term backup for other energy sources. In addition, David Novelli and Kurt Kraus studied the use of solar heat to enhance biological production of hydrogen and methane fuels as petroleum substitutes. The laboratory's knowledge of surface physics and semiconductors eventually led to investigations of ways to improve photovoltaic cells by Richard Wood and associates in the Solid State Division as part of the laboratory's modest solar program.

Management Transition

The laboratory's 1971 ventures into nonnuclear energy research did little to ease its fiscal woes. Successive annual budget reductions in its nuclear energy programs forced corresponding reductions in staff and continuous efforts to lower overhead. As one cost-cutting measure, the laboratory closed its food service canteens scattered about the complex for employee convenience and replaced them with vending machines.

Typical of his management style, Weinberg appointed long-range planners to identify supplemental laboratory missions. Commenting that he felt at times "like a man with a canoe paddle trying to change the course of an ocean liner," David Rose, the first long-range planner, soon picked up his administrative paddles and returned to MIT. Robert Livingston succeeded Rose as head of the program planning and analysis group, which included Calvin Burwell and Frank Plasil. Squarely facing the transition in ORNL missions, this group proposed a staff education program to retrain fission specialists in broader energy and environmental issues.

Musing on this proposal, Weinberg recognized the dilemma of having experts trained in one field while funding opportunities were becoming more prevalent in other fields. He noted that a similar redirection had marked the experience of Manhattan Project personnel during and after World War II. Wigner, a chemical engineer, switched to nuclear physics. Cosmic-ray specialist Ernest Wollan became a health physicist and neutron diffraction expert, and biochemist Kurt Kraus became highly skilled in plutonium chemistry. Weinberg himself had started his career as a biophysicist, only to become a reactor physicist.

"Enrico Fermi once told me that he made a practice throughout his scientific career of changing fields every five years," Weinberg recalled. He added that, although "there are few Fermis, I think we all easily recognize that the spirit of his advice can well be helpful."

In an effort to enhance the institution's viability and flexibility, in 1972 the laboratory initiated a school of environmental impacts aimed at producing physical scientists conversant with biology and ecology. This effort stalled, however, because most members of the school were laid off during the massive reduction in force of 1973. Taking cues from his own observations about the laboratory's future, Weinberg, after a quarter century of service at Oak Ridge, also embarked on a new career.

The longtime ORNL director joined Herbert MacPherson and William Baker, president of Bell Laboratories, to form a think tank dedicated to coherent long-range energy planning. With support from AEC

Floyd Culler acted as laboratory director in 1973 and later became president of the Electric Power Research Institute.

and John Sawhill of the Federal Energy Office, they created the Institute for Energy Analysis in late 1973. The Oak Ridge Associated Universities served as the institute's contract operator. It opened in January 1974 with Herbert MacPherson as director because Weinberg had been called to Washington to lend his expertise to resolving the national energy crisis.

Back on the Oak Ridge reservation, Floyd Culler served as acting director of the laboratory throughout 1973. A chemical engineer with a degree from Johns Hopkins University, Culler had worked at the Y-12 plant during the war and joined the laboratory in 1947, rising through the ranks to become a world-renowned expert on chemical reprocessing of nuclear fuels. He directed the Chemical Technology Division and served as assistant director before succeeding MacPherson as deputy to Weinberg in 1970.

Described as a "muddy boots type," Culler received acclaim at the fourth Geneva conference on atomic energy in 1971 for objecting to plans by other nations to store liquid nuclear wastes in tanks. He contended that bequeathing radioactive wastes to future generations without providing a permanent, safe disposal system posed serious political and moral questions.

Culler's year as ORNL director resembled a roller coaster ride, which he later described as a "year of many transitions." In January 1973, Milton Shaw, chief of AEC reactor development programs, mandated a quick end to the laboratory's molten-salt reactor studies. This decision precipitated what Culler described as the "largest and most painful reduction of employment level at the Laboratory in its history." It also undermined the morale of the fewer than 3,800 personnel who remained at the laboratory.

The highlight of Culler's year was the laboratory's participation in the national energy strategy. In March 1973, President Nixon appointed Dixie Lee Ray, a marine biologist, as AEC chair to replace James Schlesinger, who became secretary of defense. When the president asked Ray to review energy research and recommend an integrated national policy, she called on the national laboratories to assist in undertaking these urgent studies. Ray's report, titled *The Nation's Energy Future*, advocated energy conservation to reduce demand as well as research into new technologies and strategies, including nuclear power, to increase supplies. The report's ultimate goal was to make the nation independent of imported fuels by 1980.

The turnaround for ORNL programs came on the heels of the Yom Kippur War in the Middle East and the related Arab oil embargo of October 1973. As disgruntled Americans lined up at filling stations to purchase gasoline, Nixon established the Federal Energy Office. With William Simon as director and John Sawhill as deputy director, the office was responsible for allocating scarce oil and gas supplies during the emergency and for planning long-range solutions to the nation's energy problems.

At Sawhill's request, Weinberg went to the White House to head the Office of Energy Research and Development. Because Nixon did not appoint a presidential science advisor as had Presidents Eisenhower, Kennedy, and Johnson, Weinberg became science's delegate to the White House during the late Nixon and early Ford administrations.

Floyd Culler noted that the oil embargo and energy crisis made the laboratory "whole again" by the end of 1973. Reacting to this crisis, Congress pumped new funding into energy research and even approved a modest resumption of molten-salt breeder studies at the laboratory. "Throughout ORNL's evolution, its central theme has continued to be the development of safe, clean, abundant economic energy systems," Culler said at the end of the year. "The Laboratory is now in a uniquely strong position to undertake a multimodal attack on the nation's energy problems."

In December 1973, President Nixon proposed a reorganization of the federal energy agencies. As part of this effort, he divided AEC into two new agencies. AEC responsibilities for energy research and development went to the Energy Research and Development Administration, while AEC regulatory responsibilities were assumed by the Nuclear Regulatory Commission.

With this new administrative structure in place, Eugene Wigner recommended an ORNL reorganization paralleling the division of AEC. He urged that Weinberg be returned to ORNL to manage its energy research and development programs and that Culler be assigned responsibility for the laboratory's safety and environmental programs. "Alvin and Floyd Culler have collaborated for several years," Wigner asserted, and "they understand, like, and respect each other." As a result, he said, "conflicts are most unlikely to arise."

Wigner's recommendation was not accepted. Weinberg served the White House until formation of the Energy Research and Development Administration in late 1974 and then became director of the Institute for Energy Analysis in Oak Ridge. Culler stayed at the laboratory first as acting director and then as deputy director under Herman Postma until 1977, when he became president of the Electric Power Research Institute.

The laboratory's and AEC's transitions thus were completed by 1974. Headed successively from 1971 to 1974 by a transuranic scientist, an economist, and a marine biologist, AEC was divided into two new organizations in 1974. Changes at the laboratory were no less dramatic during these years. Managed successively by a fission scientist, a chemical processing specialist, and, in 1974, a fusion energy professional, it transcended its nuclear fission heritage to become a national laboratory embracing all forms of energy.

Life at the laboratory may have become more tumultuous during the 1970s, but changes in its workplace were no more—or less—than a reflection of dramatic changes in American society. Although isolated in the serene hills of East Tennessee, the laboratory could not avoid being caught in the vortex of a changed energy world. Its future would depend on how well it could respond to the new world "energy" order that suddenly burst on the scene in the aftermath of the Arab oil embargo of 1973 and the ensuing energy crisis.

Chapter 7

Energy Technologies

"After five years of steady decline, much personal distress, and a deep sense of frustration that obvious national problems were not being attacked," laboratory director Herman Postma said, "1974 is the year in which we perceive an end to such dismay." Warnings of energy shortages, Postma added, "finally hit home as the Arab oil embargo began and people had to wait in gas lines." The 1974 energy crisis and Postma's appointment as director during the same year had far-reaching implications for the laboratory.

Postma had joined the laboratory's Thermonuclear Division in 1959 and became division director in 1968. He was the first ORNL director without direct Manhattan Project experience. In a broader context, his ascent symbolized the arrival of a new generation of scientists—the "young Turks." These youthful scientists displayed as much interest in bioreactors, coal reactors, and fusion reactors as the laboratory's earlier researchers—now the "gray eagles"—had exhibited in nuclear reactors.

Responding to the demands of the younger scientists, Postma launched several management initiatives. Drawing on his professional management training, he initiated attitude surveys, performance evaluations, and other modern management techniques. Adhering more strictly than Weinberg to organizational structure and procedure, he strengthened the administrative role of his associate directors and divested himself of the dual roles Weinberg had filled as both laboratory director and chief of the Director's Division. Postma replaced the Director's Division with central management offices, under Frank Bruce, associate director for administration. Postma also supported creation of dual career ladders—one for scientists and technicians and another for managers. Earlier career ladders required ORNL scientists to become managers in order to obtain higher salaries.

Although the Turks and eagles may have disagreed about the laboratory's research agenda and its approach to management, both groups were pleased by a broad exploratory studies initiative begun in 1974. Dubbed the "seed money program," it aimed to encourage creative science.

"Scientific advances are made by individuals in the privacy of their own minds," observed Alex Zucker in explaining the seed money rationale. "It is one of the functions of a scientific laboratory," he continued, "to discover the unexpected, to develop new ideas, and to explore in an unfettered way areas that may not show much promise to the casual observer."

ORNL overhead funds were used to "seed" research proposals that review committees considered promising, especially initiatives that committee members thought had potential for acquiring additional funding from other federal agencies. Loucas Christophorou's study of the breakdown of insulating gases, David Novelli's amino acid research, and Elizabeth Peelle's socioeconomic analysis of power plant effects on neighboring communities were three successful seed money projects funded in 1974.

By 1977, funding had increased to one million dollars, covering start-up costs for fifteen proposals. The program remains in place today, and the laboratory's eagles and Turks, as well as the hawks of the 1990s, all view it as one of management's most successful initiatives.

What's in a Name

To Postma's surprise, in late 1974 he found himself with a new job title. No longer head of Oak Ridge National Laboratory, he became the director of Holifield National Laboratory instead—same job, same place, different title.

Aides to the congressional committees on atomic energy and government operations had memorialized their retiring chairman by renaming the laboratory after Representative Chet Holifield of California. Done without consulting Oak Ridge community leaders or ORNL

Howard Adler directed the laboratory's Biology Division from 1969 to 1975.

officials, the name change met local disapproval, although Holifield was a respected friend of Oak Ridge. "I recognize the role Holifield's played," admitted Howard Adler, director of the Biology Division, "but the name ORNL has worldwide significance and recognition that can't be tossed aside lightly."

Responding to this concern, Senator Howard Baker, Representative Marilyn Lloyd, and other members of the Tennessee congressional delegation sought to restore the name Oak Ridge. In the interim, Postma and laboratory management used Holifield National Laboratory for official government business and the familiar Oak Ridge nomenclature in scientific circles.

This conundrum ended late in 1975, when Congress reinstated the title Oak Ridge National Laboratory and named the national heavy-ion research center, a 150-foot tower under construction for the laboratory's giant accelerator, the Holifield Heavy Ion Research Facility.

More challenging than the name game was the laboratory's response to the energy crises of the 1970s. To address the fuel and heating shortages of the winter of 1974, Postma appointed Edward Witkowski and Charles Murphy as ORNL energy coordinators. Lights were dimmed and thermostats were lowered in buildings throughout the complex, and gasoline was rationed for the laboratory's fleet of vehicles. Taking these sacrifices in stride, laboratory employees donned sweaters and joined car pools to get to work. In total, emergency conservation curbed ORNL energy use by 7 percent in 1974.

Congress responded to the energy crisis by boosting the national budget for energy research, a move that helped warm and brighten (at least symbolically) ORNL's cold, dim corridors. Equally important, the energy crisis fueled congressional discontent with AEC, which had already been under fire over questions about how well it was fulfilling its safety oversight responsibilities in nuclear energy.

In 1974, Congress, following the suggestion of Nixon administration officials, voted to divide AEC into two separate agencies: the Energy Research and Development Administration (ERDA), which would serve as the federal government's energy research arm, and the Nuclear Regulatory Commission (NRC), which, as the name implies, would be responsible for regulating and ensuring the safety of the nation's nuclear energy industry.

Ending twenty-eight years of service, AEC closed at the end of 1974. Among AEC staff locking the commission's doors for the last time was Alvin Trivelpiece, later to succeed Postma as ORNL director.

ERDA absorbed the AEC laboratories, plus the Bureau of Mines' coal research centers and other federal laboratories with energy-related missions. In all, it inherited fifty-seven laboratories, research centers, and contractors—with nearly a hundred thousand employees—whom it was eager to put to work on the nation's urgent energy problems. ORNL became one of many ERDA laboratories, although its reactor safety and environmental programs also supported NRC licensing and regulatory activities.

Because no definition of laboratory roles and their relationships to other ERDA responsibilities was in place in 1974, questions about the laboratories' organization, planning, and accounting systems arose. The ERDA director, former Air Force Secretary Robert Seamans, formed a committee of advisors, including Herman Postma, to help plan the reorganization. Postma soon learned that ERDA would demand rapid applications of technology to improve the national energy posture. An ERDA official warned Postma and other laboratory directors: "If you are not working on energy projects having a good chance of being in the Sears and Roebuck catalog in five years, then you are working for the wrong agency."

ERDA's sense of urgency propelled the laboratory into a broad range of energy-related research endeavors that some wag dubbed "coconuke"—conservation, coal, and nuclear energy. At Oak Ridge, ERDA added fossil fuel and energy conservation programs to the laboratory's traditional nuclear fission and fusion energy missions—an effort that fit nicely into the broad research agenda of the young Turks.

As part of its response to the expanded mandate, the laboratory formed an Energy Division in 1974 reporting to Murray Rosenthal, associate director for Advanced Energy Systems. Samuel Beall of the Reactor Division served as the Energy Division's first director; he was followed a year later by William Fulkerson. Beall's successor at the Reactor Division was Gordon Fee, later a vice-president of Martin Marietta Energy Systems.

The new Energy Division absorbed the environmental impact reports group, the NSF environmental program, an urban-oriented social science research group, and nonnuclear studies from the Reactor Division under one administrative umbrella.

The Energy Division, briefly stated, sought to tie energy research and conservation to broad questions of social and environmental impacts. In effect, the laboratory had acknowledged within its administrative framework that energy research could no longer be confined to narrow technical issues.

Energy Conservation

Recognizing that the nation's energy posture could be improved by reducing consumption of existing energy resources and putting wasted energy to use, the laboratory joined ERDA's national conservation program. Through many small enhancements in energy conservation, the laboratory and ERDA expected in the aggregate to reduce national energy use by several percentage points annually.

Some conservation research stemmed from the laboratory's earlier studies of the potential environmental effects of nuclear power plants. Having observed the discharge of waste heat from these plants into the water and air, ORNL researchers proposed putting that waste to use by warming both greenhouses to grow plants and ponds to raise fish for food. As an outgrowth of ORNL recommendations, TVA and other electric power utilities undertook experiments with greenhouses and related heat-use facilities in the design, construction, and operation of their nuclear plants during the 1970s.

The laboratory proposed similar uses for waste heat, called cogeneration, for a modular integrated utility system it blueprinted for the Department of Housing and Urban Development (HUD). In this design by John Moyers and others for small communities, heat—produced as a byproduct of electricity from a generating plant—would supply space heating and hot water.

With funding from the Department of Housing and Urban Development (HUD), ERDA, and NSF, six ORNL divisions (including the Energy Division) launched a comprehensive set of programs to foster energy conservation in 1974. Moreover, because of strict personnel ceilings, ERDA asked the laboratory to act as its program manager for conservation efforts throughout the energy agency's sprawling federal network.

For ERDA, the laboratory planned conservation programs, awarded subcontracts for research and engineering, and monitored and reviewed the work. Many of these responsibilities were carried out by the laboratory's residential conservation program headed by Merl Baker and Roger Carlsmith. The program supported studies of improved home insulation, tighter mobile home design, advanced heating and cooling systems, and energy-efficient home appliances.

When ERDA asked the laboratory to assess how much energy could be saved by better insulating homes and businesses, Ralph Donnelly, Victor Tennery, and colleagues undertook a study that, in 1976, reported that improved insulation was crucial to national energy conservation. The laboratory emerged as ERDA's prime resource for developing thermal insulation standards, later adopted by ERDA, the Department of Commerce, and building trade associations. These standards helped generate substantial and continuing savings for homeowners while paring national energy consumption. Retrofitting existing buildings to save energy followed when electric utilities, such as TVA, financed improved home insulation, heat pumps, and other ORNL-developed energy conservation measures in existing structures.

Manufactured homes promised energy savings that likely would exceed savings in more conventional structures. ORNL studies, led by John Moyers and John Wilson, sought to determine the full range of potential savings. "Mobile homes are produced in factories," Moyers pointed out, "and should be more susceptible to quality control, unified system design, and engineering than custom-built homes."

Relying on data obtained from a mobile home equipped at ORNL with instruments to measure its power use and seasonal temperature fluctuations, researchers proposed tighter insulation and storm window standards subsequently adopted by the American National Standards Institute and HUD to upgrade mobile home energy efficiency. Those who purchased new mobile homes, often recently married couples or retirees with limited incomes, enjoyed reduced energy costs, and the nation as a whole cut its energy consumption.

Harry Fischer's annual cycle concept may have been the most publicized ORNL energy conservation endeavor. A retiree with years of experience in energy engineering, Fischer dropped by the laboratory in 1974 to tell Samuel Beall, new director of the Energy Division, that he knew how to provide home heating and cooling at half the cost of systems then in use. His annual cycle system used a heat pump that extracted heat during winter from a large insulated tank of water, changing the water into ice for summer cooling.

Returning home to Maryland, Fischer discussed his concept with his neighbor, Secretary of Interior Rogers Morton, who offered support if Fischer and the laboratory could produce a working model within three months. They had it operating in two. Fischer also met John Gibbons, formerly with ORNL and now heading the University of Tennessee's Energy, Environment, and Resources Center. Gibbons, who was overseeing the university-sponsored construction of houses using solar and conventional heat near Knoxville, agreed to construct a third home using Fischer's annual cycling system next to the others. Jointly managed by the university, the laboratory, TVA, and ERDA, the houses were completed in a year. ERDA Director Seamans personally inspected them to highlight the fast action that had been demanded.

As Fischer predicted, the annual cycling system house could be heated and cooled at half the energy costs of conventional systems; in fact, the system earned ultimate energy savings one month when Fisher's team forgot to pay the electric bill and the local utility company cut off its power. However, few ever adopted Fischer's system, largely because of its high initial capital costs and potential maintenance problems.

Another ORNL conservation project that received broad media attention was its bioconversion experiment, called ANFLOW. In 1972, Congress mandated secondary sewage treatment for all communities. The laboratory estimated the new systems would double the energy used for sewage treatment, so it decided to explore technologies that might reduce energy consumption and costs. Alicia Compere and William Griffith, working with John Googin at the Y-12 plant, devised a bioreactor, known as ANFLOW, to explore its energy-saving possibilities in treating sewage.

Conventionally activated sludge sewage treatment used oxygen-seeking aerobic bacteria to digest wastes. In contrast, the ANFLOW system used anaerobic microorganisms that did not require oxygen. This process eliminated the need for energy-consuming pump aerators. Moreover, the ANFLOW system could produce methane gas, useful as fuel, from sewage, and recover valuable chemicals from industrial wastes for reuse.

On its own, the laboratory built an experimental ANFLOW bioreactor, and in 1976, it contracted with the Norton Company to build a pilot ANFLOW bioreactor to be installed at an Oak Ridge municipal sewage treatment plant. The ANFLOW bioreactor pumped sewage through a fifteen-foot cylinder packed with gelatin-coated particles to which microorganisms attached themselves. The packing, made of crushed stone or ceramics, facilitated the waste flows and provided additional surfaces for the microorganisms, which thrived and reproduced while consuming wastes.

The Chemical Technology Division's Richard Genung, Charles Hancher, and Wesley Shumate managed the ANFLOW program, and in 1978 awarded a subcontract for the design of a larger demonstration plant, installed as part of the Knoxville sewage treatment system. Potato processing, meat packing, and other industries expressed interest in this waste-treatment method.

Research on the use of organisms to treat waste, however, has proceeded slowly. Moreover, municipalities seldom build new sewage treatment plants; they are capital-intensive, time-consuming projects that may require a decade or more to negotiate and construct. Therefore, energy savings derived from more efficient sewage treatment would be a long time coming. Despite these obstacles, work on ANFLOW has encouraged broader ORNL investigations into potential biological solutions to waste-disposal problems.

In contrast to long-lived sewage systems, homeowners replace several electric appliances each decade. Believing that aggregate energy savings could be substantial, ORNL researchers launched detailed studies of ways to improve the efficiency of heat pumps, refrigerators, furnaces, water heaters, and ovens. Eric Hirst, Robert Hoskins, and colleagues in the Energy Division gained wide acclaim for computer modeling of home appliances to identify opportunities for greater energy efficiency. Their computer analysis of refrigerator designs, for example, indicated that energy use for these appliances could be halved through installing better insulation, adding an antisweat heater switch, improving compressor efficiency, and increasing condenser and evaporator surface areas.

ORNL energy-saving recommendations for home appliances were incorporated into the design standards of the American Society of Heating, Refrigerating, and Air-Conditioning Engineers and also into experimental appliances designed by subcontractors under the management of Virgil Haynes at the laboratory. Out of this applied research came more efficient appliances, notably a heat-pump water heater and refrigerator, which were soon manufactured for commercial markets. By the 1990s, most American homes had at least one appliance that was more energy-efficient as a result of the laboratory's conservation research.

Fossil Energy

With nearly half of the world's known coal reserves, the United States has been called the "Saudi Arabia of coal." In the face of dwindling domestic petroleum supplies, scarce natural gas reserves, and the uncertainty and escalating price of oil imports, it seemed logical in the 1970s to supplement petroleum with fuels produced from coal.

Scientists had long known that applying heat and pressure to coal could produce liquids, gases, and chemically altered solids for fuel. Efforts to turn scientific theories and blueprints into commercial ventures, however, had been minimal. Then, in 1975, ERDA announced its goal of producing a million barrels of synthetic oil from coal daily by 1985. To produce that much synthetic fuel would require as many as twenty plants, so ERDA contracted with industry to plan and design a series of pilot demonstration plants. ERDA's Oak Ridge Operations office managed the contracts and obtained research support from the laboratory.

In response to this major federal initiative, Murray Rosenthal announced an interagency agreement with the Office of Coal Research that brought ORNL into fossil energy research. This agreement culminated in a coal technology program headed by Jere Nichols, later renamed the fossil energy program under Eugene McNeese, that was budgeted at $20 million annually. It included fundamental studies of the structure of coal, the carcinogenic properties of coal conversion products, a hydrocarbon reactor, and a potassium boiler to improve the efficiency of producing electricity by burning fossil fuels. Under this program, the laboratory exchanged personnel and collaborated with the Bureau of Mines' coal laboratories at Bruceton, Pennsylvania; Morgantown, West Virginia; and Laramie, Wyoming.

Planning to fund industrial pilot and demonstration plants that used synthetic refined coal and hydrocarbonization processes, ERDA assigned the laboratory a major role in evaluating the progress of this broad-ranging initiative. For one project, Henry Cochran and colleagues in the Chemistry and Chemical Technology divisions built a model hydrocarbon reactor that mixed finely ground coal with hydrogen under high pressure and heat to form synthetic oil, plus a substitute for natural gas and a cokelike solid fuel. Modeling experiments identified the optimal combination of pressure and heat for fuel production. Related projects conducted by Richard Genung, John Mrochek, and their colleagues included studies of coal thermal conductivity and recovery of aluminum and minerals from fly ash.

A bioprocessing group, led by Charles Scott of the Chemical Technology Division, launched a series of studies of bioreactors. The dual goal was to concentrate and isolate trace metals and to produce liquid and gaseous fuels organically. In bioreactors resembling those in the ANFLOW sewage treatment project, microorganisms adhering to fluidized particles in columns could digest toxic compounds from the wastes of coal-conversion processes, converting them to harmless substances.

Researcher Chet Francis in the Environmental Sciences Division demonstrated that simple garden soil bacteria in bioreactors could remove nitrates and trace metals from industrial wastes effluents. As a result, the laboratory built a pilot bioreactor used by the Portsmouth, Ohio, gaseous diffusion plant to treat nitrate wastes, and the Y-12 plant used Francis's design for a full-scale plant to treat nitric acid wastes.

The laboratory also looked for ways to reduce air pollution caused by the burning of coal. In the Engineering Technology Division, John Jones's team developed a fluidized-bed coal reactor connected with a closed-cycle gas turbine for power generation. Aiming at making high-sulfur Appalachian coal more environmentally acceptable, the system fed coal and limestone particles into a furnace where jets of preheated air agitated them, igniting the coal and thus providing the heat needed to combine the limestone with sulfur dioxide to form nontoxic gypsum. ERDA sponsored the construction at the Y-12 plant of a prototype to prove that Appalachian coal could be burned cleanly during power generation.

Eugene Hise and Alan Holman devised another method of cleaning sulfur from coal. Because sulfur-bearing iron pyrites and ash-forming minerals are weakly attracted by magnetic fields and coal particles are mildly repelled, they devised a system for magnetically cleaning coal, using a superconducting solenoid to provide a magnetic field of the required shape and force.

In another coal-related research initiative, NSF funded a regional evaluation of the economics of strip-mine reclamation in Appalachia. Robert Honea and Richard Durfee headed a team in 1975 that used satellite imagery, census data, and regional-scale models to analyze strip mining. Focusing on mining in the New River basin north of Oak Ridge, the study took images from space satellites to classify land cover types, which were then verified with aerial photographs. Researchers could examine strip-mining effects during every overhead pass of the satellite, enabling them to obtain a better picture as the mining unfolded instead of just a snapshot of the impacts once the mining was complete.

In 1975, ERDA Director Seamans broke ground for the Environmental Sciences Laboratory in Oak Ridge. A two-unit structure, it became ERDA's first programmatic laboratory. ORNL's first major laboratory and office expansion since the 1960s, Environmental Sciences was located at the west end of the complex near the Aquatic Ecology Laboratory. The main building was connected by walkways to greenhouses, animal and insect facilities, and chambers for controlled-environment experiments.

Chester Richmond, who succeeded James Liverman and John Totter as the associate director for Biomedical and Environmental Sciences, in 1976 implemented a life sciences program to support coal-conversion technologies. Working closely with the Environmental Protection Agency (EPA), the program, led by ecologist Carl Gehrs of the Environmental Sciences Division, examined the chemical and physical characteristics of coal liquids, their biological and health effects, and their transport through ecosystems.

From this effort came funding for examining mutagenesis (in the Biology Division), ecological toxicology (in the Environmental Sciences Division), health risk effects (in the Health and Safety Research Division), and coal-liquid constituent identification (in the Analytical Chemistry Division). These initiatives enabled the laboratory to prove that coal-conversion liquids and effluents could be toxic. It also provided information to guide changes in coal chemical processing that would create less toxic products.

Fusion and Fission Energy

Under ERDA, ORNL fusion energy research expanded more rapidly than fission research. Although fusion research could not enhance the nation's short-range energy posture, ERDA gave the program substantial support in the hope that it would ultimately provide a long-range solution to the nation's energy problems. But ERDA's research agenda held little promise for the laboratory's remaining gray eagles, whose careers had focused on fission research. With the end of the molten-salt reactor and modest support for high-temperature gas-cooled reactor research, the research agenda of the laboratory's Manhattan-era researchers had been reduced to the Clinch River breeder reactor technology and related fuel reprocessing for plutonium recovery.

The impurities studies experiments (ISXs) were a focus of ORNL fusion research during the 1970s.

Under John Clarke, Postma's successor as chief of fusion energy research, successful testing of the ORMAK and ELMO Bumpy Torus devices continued into the 1970s. The laboratory also built impurities study experimental devices (ISXs) to illuminate the behavior of impurities inside fusion reactor plasmas. Researchers, led by Stan Milora and Christopher Foster, developed a pellet injection method, firing frozen hydrogen pellets into fusion plasmas to maintain the plasma densities. This refueling technology was subsequently adopted for tokamaks in Europe and the United States.

A major research challenge of the 1970s was the action of fusion plasma when it escaped the magnetic field and met the wall of its containment vessel. Would it sputter impurities from the wall back into the plasma and poison the reaction? To study this and related questions, the laboratory formed a "first wall interactions" group led by William Appleton, James Roberto, Robert Clausing, Robert Langley, and Peter Mioduszewski. This group cooperated with similar fusion research groups at other laboratories.

Other fusion research advances during the ERDA years included the neutral beam technology developed by William Morgan's team to heat plasma inside a fusion device. This technology helped Oak Ridge's ORMAK and Princeton's tokamak achieve record temperatures that approached the break-even point needed for a self-sustaining reaction.

Investigations of huge superconducting magnets for containing fusion plasmas began under Hugh Long, Martin Lubell, Fred Walstrom, and William Fietz, leading to the selection of the laboratory in 1977 to build the Large Coil Test Facility. Managed by Paul Haubenreich, this facility would test supercold magnets, weighing forty tons each, which were manufactured both in the United States and abroad.

While fusion energy research prospered, the laboratory built no new nuclear reactors during the 1970s. In 1976, the laboratory changed the name of the Reactor Division to Engineering Technology because its work no longer concerned overall reactor design; instead, it focused on the development of engineering systems for both nuclear and non-nuclear facilities. The nuclear safety program for NRC continued, however, under Fred Mynatt.

After 1976, the laboratory's nuclear energy research centered largely on the Clinch River breeder reactor project and plans to reprocess its fuel. Design of the steam generator and heat exchangers for the Clinch River reactor was undertaken by ORNL metallurgists led by Peter Patriarca, who investigated thermal stress and creep in the materials to be used in these systems.

The laboratory also specialized in devising materials for breeder and fusion reactors that would withstand radiation damage. James Weir developed a theory to explain how heated steels swelled and became embrittled during neutron bombardment in reactors, and researchers James Stiegler, Everett Bloom, Arthur Rowcliffe, and others developed improved stainless steel alloys doped with silicon and titanium. Despite these advances, support for Clinch River breeder programs faltered after the election of President Jimmy Carter, who opposed the project on the grounds that the United States could achieve energy independence through strategies and technologies that would be less costly and pose less risk to the environment and world peace.

Splendid Crowding

The years of urgent energy research under ERDA were a time of expansion for the laboratory. By 1977, it had acquired lead responsibility for five major ERDA programs and had become involved with the full complement of the nation's energy programs. In addition, it had undertaken work for eleven other agencies, amounting to $35 million in funding annually, and it was subcontracting six times the amount of outside work it had supported in 1974. The number of ORNL personnel rose to more than 5,000, performing and supporting about 700 scientific and technical projects. The laboratory also hosted 1,250 guest researchers and more than 25,000 visitors annually.

In particular, emphasis by ERDA on developing nonnuclear advanced energy systems proved a boon for materials sciences at the laboratory. Limited previously to investigations related to the fission and fusion energy programs, under ERDA the materials sciences advanced into studies of many materials. This was especially pertinent to the Solid State Division under Michael Wilkinson and the Metals and Ceramics Division under James Weir, which experienced significant program expansions.

Although the Environmental Sciences Laboratory and the Holifield Heavy Ion Research Facility were under construction in 1977, ORNL had not added significant space to its complex since the 1960s. Existing work space was reduced even more by the addition of minicomputers and copying machines during the 1970s. The stereotype of scientists cogitating in splendid isolation was far from true at ORNL in 1977. In fact, conducting research there had become a close-quartered affair.

"The fact is that programs grow faster than buildings can get built or that money can be found for that purpose," lamented Postma. "In practice, the only justification for new buildings is to alleviate crowded conditions that already exist rather than rationally anticipating projected needs," he elaborated. "Thus, in the future there will be more crowding at the Laboratory, more sharing of offices, and far greater need for understanding and cooperation by all members of the Laboratory."

The problem of overcrowding abated unexpectedly in 1977 when

newly elected Presider␣ Jimmy Carter and his Department of Energy
(DOE) adopted person␣ ␣l ceilings that capped the number of ORNL
employees. After four␣ ␣ars of nearly nonstop additional hiring, the
laboratory's personnel␣ ␣fices suddenly became tranquil and quiet.

President Carter w␣ ked to the White House in January 1977 in the
midst of one of the twentieth century's coldest winters. At the time, the
effects of the 1973 oil crisis still rippled through the national economy.
Unprecedented cold temperatures generated unanticipated demands
for energy supplies, placing additional stress on a national energy system that had not fully adjusted to the post-OPEC energy world.

The result was another energy crisis, although not nearly as severe
as the paralyzing events that had gripped the nation four years before.
Nevertheless, during the oil and natural gas shortage of 1977, the laboratory narrowly avoided a complete shutdown for lack of heat only because
the K-25 plant shared its on-site oil reserves during the emergency.

Calling for the "moral equivalent of war" on energy problems, President Carter in the spring of 1977 requested public sacrifices for the sake
of regaining control of the nation's energy future. To manage the battle,
he proposed establishing a cabinet-level Department of Energy. Approved by Congress in August 1977, the new Department of Energy
absorbed the functions of ERDA, the Federal Energy Administration,
and the Federal Power Commission, plus energy programs from other
federal agencies.

Carter appointed James Schlesinger, former AEC chairman and
secretary of defense, the nation's first energy secretary. In addition, the
president announced his opposition to the Clinch River breeder reactor project and stopped the reprocessing of nuclear fuel. These decisions clouded the future of nuclear energy, which, in turn, placed the
future of the laboratory's nuclear divisions on an uncertain path with no
clear signposts. For the gray eagles, the breeder was the future. If the
project was abandoned, where could they turn to pursue their lives' work?

Stability amid Transition

The transition from ERDA to DOE proved difficult. ERDA administra-
tor and assistant administrators resigned before DOE became functional

in October 1977, leaving agency program direction unclear. "Whereas we perceive uncertainty and lack of clear direction in Washington, the realities at the Laboratory are quite different," observed Alex Zucker during this transition. "Our programs are productive, our staff is busy. Stability rather than uncertainty characterizes our work; and, if we work now in new areas, we are doing it with the old élan."

Secretary Schlesinger revised the system for managing DOE's eight multiprogram laboratories, thirty-two specialized laboratories, and sixteen nuclear materials and weapons laboratories. For their institutional needs, the laboratories were to report to assistant secretaries in Washington instead of regional operations officers.

Invited to Washington to advise Schlesinger on basic research needs, Postma declared that integrating energy development into a single department at last recognized that energy was as important as labor, agriculture, and defense. "There will be studies galore to evaluate everything," Postma predicted. He was confident that the laboratory would prosper despite the "turbulence represented by the changing political and programmatic winds in Washington."

During 1978, the transition to DOE was completed. Believing that national laboratories had reached optimum size, the Carter administration sought to work more directly with industry, expanding the role of national laboratories as program and subcontract managers. It designated national laboratories as centers of excellence in special fields and imposed ceilings on the number of personnel. Oak Ridge was made the lead laboratory for coal technology and fuel reprocessing, and was told that its staff could not exceed 5,165 personnel for 1979.

The Carter administration proved more interested in energy conservation and "soft," renweable energy than in nuclear energy. Taking its cues from Washington, ORNL began to emphasize small programs in geothermal and solar energy initiated under ERDA. The Environmental Sciences Division also initiated intensive study of wood and herbaceous biomass—fast-growing trees and grasses that could be converted to a renewable energy resource.

John Michel managed the laboratory's research on geothermal energy using hot water formed within the earth. This initiative encompassed research in the Chemistry Division on scaling and brine chemistry, in the Metals and Ceramics Division on corrosion, and in the

Engineering Technology and Energy divisions on cold-vapor, low-temperature heat cycles. The collective goal of this technical research was to upgrade the efficiency of producing electricity with geothermal energy.

A related research program studied ways to improve heat exchangers to capture the oceans' thermal energy. Rather than burning the rocks and the seas with nuclear energy—a dream of the 1960s—this research sought to extract low-level energy from the earth and ocean in kinder and gentler ways.

The laboratory's solar energy research was circumscribed by formation of a special DOE laboratory, the Solar Energy Research Institute in Colorado (now called the National Renewable Energy Laboratory). Robert Pearlstein became coordinator of Oak Ridge's small solar program, which included research in the Chemistry and Solid State divisions. Eli Greenbaum and associates in the Chemistry and Chemical Technology divisions investigated the production of hydrogen from water by using green plant materials to capture and convert the sun's energy catalytically, while the Solid State Division program under Richard Wood investigated improved photovoltaic solar cells for converting sunlight directly into power.

With initial funding from the seed money program, John Cleland's team in the Solid State Division developed a new method of doping silicon to produce the semiconductors used in solar cells. Instead of using chemical doping methods, a silicon isotope in samples inserted in the bulk shielding reactor was transmuted into phosphorus through interactions with neutrons. This process provided uniform distribution of phosphorus in the silicon, thereby improving the efficiency of solar cells fabricated from this material.

In a related development, the Solid State Division in 1978 used lasers in preparing silicon for solar cell fabrication. To provide good distribution within the silicon, ions of a dopant such as boron were deposited on a silicon surface or implanted into the surface using an ion accelerator. Lasers were used for annealing semiconductors to remove crystal imperfections introduced in the implantation process. This combination of ion implantation doping and laser annealing, initiated by Rosa Young and C. W. White, spurred fundamental and applied studies in the processing of solids.

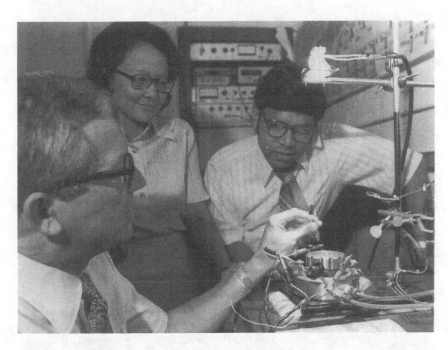

Richard Wood, Rosa Young, and Jagdish Narayan test the solar cells made with laser-induced diffusion at the laboratory.

The surface modification and characterization facility, started by William Appleton and now headed by White, became the focal point for these studies. Housed initially in the old fan house of the graphite reactor, the center became a DOE user facility hosting university researchers and industry collaborators.

Almost a Mecca

To observe firsthand the laboratory's research achievements and to soothe the ill feelings generated by his decision to oppose the Clinch River breeder reactor project, President Carter visited Oak Ridge in May 1978 at the request of Senator James Sasser. The president brought his science advisor and energy staff with him. Remembering his service as an officer in Admiral Rickover's nuclear navy, Carter declared, "Oak Ridge was almost like Mecca for us because this is where the basic work was done that, first of all, contributed to the freedom of the world and ended the war and, secondly, shifted very rapidly to peaceful use of nuclear power."

The first president to visit the laboratory while in office, Carter received a cordial welcome. In the Building 4500N lobby, Postma introduced him to Charles Scott, who outlined bioreactor experiments; to Samuel Hurst, who discussed lasers that detected single atoms among millions; and to John Jones, who described a fluidized-bed coal burner designed to cogenerate power and heat. Afterward, the president enjoyed technical presentations and a roundtable discussion with a group of scientists in the auditorium.

There, the president seemed particularly interested in Lee Berry's description of fusion research, asking how it compared with Soviet research. Berry responded that the United States may have enjoyed a slight lead in the fusion race. Sandy McLaughlin appealed to the president's environmental interests by describing ORNL research on the ecological effects of atmospheric pollutants.

Laboratory personnel greeted the president with respectful cheers, surprising local reporters who thought the president's opposition to the proposed Clinch River breeder reactor and subsequent political decision to move a centrifuge plant from Oak Ridge to Portsmouth, Ohio, would elicit a cold though polite silence, or even murmurs of discontent.

President Carter, however, was near the peak of his popularity at the time of his visit to the laboratory. Afterward, events such as the Iranian hostage crisis plagued his administration, exacerbating the national energy crisis and inevitably affecting ORNL activities.

Quick Responses

The March 1979 accident at Three Mile Island Unit 2 surprised nuclear experts at ORNL and elsewhere. Although nuclear safety research had concentrated on the risks of pipe rupture and the possibility of loss-of-coolant accidents in light water reactors, the Three Mile Island accident in Pennsylvania resulted instead from a pressure valve that stuck and inaccurate instrumentation and human error that complicated and exacerbated the emergency. Having a national reputation in the safety field, ORNL staff led by Fred Mynatt became immersed in the Three Mile Island emergency and subsequent analysis.

When the company owning the disabled reactor called Floyd Culler at the Electric Power Research Institute for help, Culler (who had just left ORNL after twenty-five years of service, including one year as acting director) contacted Postma and other laboratory officials, as did the staff of NRC. During the emergency, laboratory personnel served as consultants and on-site analysts. Specifically, seventy-five staff members performed technical and analytical research, or subsequently provided information to the committee appointed by President Carter to investigate the accident.

The laboratory helped the industry recover from the accident in many ways. An Industrial Safety and Applied Health Physics Division team led by Roy Clark monitored the radiation emissions, while Robert Brooksbank's team minimized radioactive iodine releases by adding chemicals to the cooling system and by arranging replacement of the filters used to cleanse reactor gases before their release into the atmosphere. The absence of significant iodine releases was in part a testament to their success.

The Chemical Technology Division designed systems to store the contaminated water and remove the fission products. Robert Kryter and Dwayne Frye of the Instrumentation and Controls Division supervised the installation of monitors that replaced the damaged sensing systems inside the reactors. Wilbur Shults and an analytical chemistry team analyzed samples from the accident site to assess the severity of contamination and devise cleanup strategies. Mario Fontana's engineering technology group and David Hobson's metals and ceramics team examined core cooling and debris problems, zircalloy cladding damage, and fission product releases. David Bartine's group addressed radiation and shielding issues. Joel Buchanan led the team studying the hydrogen bubble in the reactor, and David Thomas supervised an Engineering Technology Division group that simulated the accident in the thermal-hydraulic test facility.

Accident investigations and recovery activities continued for years, and the laboratory took pride in its emergency and long-term responses.

The review by Anthony Malinauskas and David Campbell of the issues surrounding radioactive iodine releases to the atmosphere for President Carter's commission and NRC proved especially fruitful.

The accident at Three Mile Island forever changed the public's attitude toward nuclear power and the high priests of science. ORNL's response, however, helped provide a sound scientific base for understanding the causes and effects of the most serious mishap in the history of the U.S. commercial nuclear industry.

Later in 1979, the nation and the laboratory became troubled by the revolution in Iran and the hostage and energy crises that ensued. Visiting Iran shortly before the revolution to discuss training Iranian technicians at ORNL, Associate Director Donald Trauger observed firsthand the political instability there. He refused, however, to describe the subsequent acute petroleum shortage as another energy crisis. After a decade of energy crises, he believed that it was time for the nation and world to accept the shortages of adequate energy supplies as a persistent and chronic problem. The word crisis suggests an unexpected situation that can be set straight by rapid, aggressive responses. Instead, Trauger suggested that "we must hurry to find solutions, but we must not become overly impatient in our quest."

ORNL energy conservation efforts accelerated during the Iranian embargo. The laboratory converted its steam power plants from natural gas and petroleum back to coal and turned to gasohol to fuel some of its vehicles. In addition, the laboratory's environmental impacts group was commandeered to analyze implementation of the Strategic Petroleum Reserve—a federally sponsored effort to store large quantities of oil that could be tapped in times of emergency. The Strategic Petroleum Reserve later would serve a valuable role in stabilizing oil prices during the Persian Gulf War of 1991.

Constancy of Change

In 1980, the laboratory found itself caught uncomfortably in the impasse between Congress and President Carter over the Clinch River breeder reactor project. Funding for the laboratory's breeder research to support the reactor and fuel reprocessing was slashed significantly—

a blow to fission research that further discouraged the laboratory's dwindling number of charter members.

"This last defeat has convinced gray eagles like myself that the rainbow we have been following for the past 30 years may indeed not have the long sought after pot of gold at the end," lamented Peter Patriarca, head of the laboratory's breeder reactor materials research program. "I feel that I and others like me have accomplished a lot in 30 years of service," he concluded, "but we really haven't achieved the ultimate and that is my disappointment."

Still, 1980 was a banner year for many ORNL programs. For the first time, the budget exceeded $300 million. Of this total, $20 million was subcontracted to universities and $60 million to industry to support research and engineering. Completed in 1979, the new Environmental Sciences Laboratory eased staff crowding. Three new user facilities opened in 1980, marking the culmination of three successful programs launched in the 1970s: the National Environmental Research Park, Holifield Heavy Ion Research Facility, and the National Center for Small-Angle Scattering Research.

The user facility concept evolved from a fundamental change at DOE. Before 1979, many ORNL personnel collaborated informally with scientists from outside ORNL. That year, DOE made it official policy that its facilities were to be opened to outside users for cooperative and proprietary research and development.

The Oak Ridge National Environmental Research Park, comprising 12,400 acres of protected land for environmental science research and education, opened in 1980 as the fifth outdoor laboratory of DOE. Nearly surrounding ORNL, it made up about a third of the Oak Ridge reservation. Here, scientists inventoried plant and animal species; monitored the dynamics behind climate and ecological change; undertook studies of contaminant transport and bioremediation; and cooperated with local, regional, and private agencies to promote science and environmental education.

Nearly twenty thousand students from kindergarten to high school visited this park annually as part of their science education programs. The Walker Branch Watershed in the park emerged as a key experimental facility for biogeochemical and hydrologic research.

One early research effort in the park tested bird and small-animal

The Joint Institute for Heavy Ion Research is an energy-efficient building for scientists visiting the Holifield Heavy Ion Research Facility.

habitat models later used by the Army Corps of Engineers to prepare environmental impact statements for construction projects. Another early research effort examined atmospheric deposition of pollutants for the National Oceanic and Atmospheric Turbulence and Diffusion Laboratory located in Oak Ridge. These efforts helped counteract the negative publicity that accompanied disclosures that portions of the reservation had been contaminated by the release of radioactive and other toxic materials into the environment, especially between the 1940s and early 1960s.

Former Congressman Chet Holifield participated in the December 1980 dedication of the Holifield Heavy Ion Research Facility named after him. "One more curiosity of the scientifically oriented human mind" is how Holifield described the awesome tower and the accelerator it housed.

Twice as powerful as any machine of its type, the accelerator in the tower was coupled with the Oak Ridge isochronous cyclotron to convert heavy ions into high-speed projectiles. Colliding with targets, these

The small-angle X-ray scattering device became part of the National Small-Angle
Scattering Facility established at the laboratory in 1978.

projectiles produced results that helped illuminate fundamental nuclear
science. Laymen were more amazed by the spin spectrometer, a clustered
array of gamma ray detectors, dubbed a "crystal ball," used to measure the energies of the gamma rays emitted by the products of the
heavy-ion collisions.

Like the environmental park, the Holifield heavy-ion facility was
designated a national DOE user facility. Over the years, it has hosted
numerous scientists from around the world. By the late 1980s, in fact,
nearly a quarter of all doctoral degrees in low-energy nuclear physics
involved work done at this facility. Oak Ridge Associated Universities
organized the UNISOR group of universities that established a unique
isotope facility at Holifield. The Physics Division under Paul Stelson
and later James Ball formed and built the Joint Institute for Heavy Ion
Research on DOE land using funding from Vanderbilt University and
the University of Tennessee. The institute was not only a model of cooperation, it was also a mostly underground, energy-efficient structure

designed by ORNL architect Hanna Shapira, which served as a visible symbol of the laboratory's commitment to energy conservation.

The National Center for Small-Angle Scattering Research, the laboratory's third user facility, opened in 1980. Small-angle neutron scattering blossomed during the 1970s as a way to explore microscopic structures with dimensions of ten to one hundred angstroms. Although two laboratories using this scientific technique existed in the United States, they were not readily available to independent researchers, and in 1977, NSF proposed to fund a center for use by scientists nationwide.

Wallace Koehler and Robert Hendricks, who had developed a small-angle X-ray scattering instrument, submitted a proposal to establish a user-oriented, small-angle scattering center at the laboratory. It called for a new small-angle neutron-scattering (SANS) facility at the high-flux isotope reactor along with access to the laboratory's existing small-angle X-ray and neutron-scattering devices. Their competitive proposal, which received NSF approval in 1978, also included computer equipment that allowed users the luxury of largely automated experimentation.

The new SANS facility, which opened in 1980 at the high-flux isotope reactor, included a position-sensitive detector designed by Casimir Borkowski and Manfred Kopp. Directed by Koehler and George Wignall, the facility compared well with the best facilities in Europe, and the center offered a combination of X-ray and neutron scattering that made the laboratory a mecca for this type of materials research.

With these new facilities, the laboratory entered the 1980s prepared for its role as a user-oriented institution that could host scientists from around the world. After a decade of energy crises and constant transition, ORNL seemed to have adjusted well to its new role as a DOE multiprogram laboratory.

During the presidential election of late 1980, however, candidate Ronald Reagan complained that DOE had not produced a single additional barrel of oil and promised to dismantle Carter's creation. By Christmas of that year, Reagan's transition team announced it had profound changes in mind for both DOE and its national laboratories.

In less than a month, they would have an opportunity to put those ideas into practice. Barely having caught its breath from a decade of whirlwind change in energy policy and direction, the laboratory was poised for yet another transition. The Reagan years were about to begin.

Chapter 8

Diversity and Sharing

In the early 1970s, ORNL began to move beyond its war-rooted preoccupation with nuclear power to research fields embracing all energy forms. By the early 1980s, that journey was complete. In the words of Associate Director Alex Zucker, Oak Ridge had become "a multiprogram research and development laboratory having a variety of energy-related missions of national importance."

Emphasis on the laboratory's multiprogram character was in part a response to the "Reagan revolution" of the 1980s, when fierce debates arose over the proper balance between the public and private sectors. The Reagan administration, in fact, proposed to abolish DOE and severely curtail the activities of the national laboratories. Energy policies, the administration stridently proclaimed, should be shaped by the private sector. If government had any role at all, it should be narrowly confined to questions of basic research.

President Reagan appointed James Edwards, a former governor of South Carolina and oral surgeon with little background in energy policy, to preside over DOE's dissolution as the nation's "last" secretary of energy. The president planned to transfer the department's residual functions to the Department of the Interior under James Watt or to the Department of Commerce under Malcolm Baldrige.

Seeking—at least rhetorically—major budget reductions in the public sector, in 1981 the Reagan administration initiated executive reviews of most federal agencies, including DOE laboratories. Kenneth Davis, deputy secretary of energy under Edwards, directed an Energy Research Advisory Board to survey the laboratories' work. Congress conducted similar studies.

Investigators distinguished between three kinds of laboratories: single-purpose specialty, exclusively weapons, and broadly diverse multiprogram. Oak Ridge, Argonne, and Brookhaven were the original multiprogram laboratories, but the list soon expanded to include more than a dozen DOE laboratories eager to join the multiprogram bandwagon.

Nevertheless, vocal criticisms of these multiprogram laboratories arose from universities, consulting firms, and industrial laboratories. Because of the laboratories' excursions during the 1970s into diverse energy research agendas, critics saw them as subsidized competition. One industrial executive, for example, charged, "When I find Oak Ridge planting trees to see if they can't grow them a little closer together and faster, which the paper companies could do; testing solar cells that there are 300 companies already set up to test; and so on, I just wonder if we haven't lost our sense of focus altogether."

Admitting the missions of national laboratories had become diffuse and perhaps "unfocused" during the 1970s, ORNL leaders asked whether more precise definitions of the roles of all laboratories—national, private, and university—would help clarify the situation and foster a healthier and more robust national research program. Truman Anderson, chief of laboratory planning and analysis, urged that national laboratories should "assume a broader role in a new partnership with industry and universities." This new partnership was to reshape ORNL activities throughout the 1980s and into the 1990s.

Program diversity enabled the laboratory to weather the intense scrutiny of 1981; so, too, did the administration's pronuclear stance, which ameliorated its initially harsh approach to government-sponsored energy programs. Commenting on the effects of Reagan's policies after his first year in office, Herman Postma, the director of the laboratory, declared, "The impacts . . . so far, while unwelcome and frequently painful, have been rather moderate overall, and certainly less severe than at many of our sister laboratories." Indeed, Postma thought the Reagan policies had some salutary effects, notably in restoring an equitable balance between basic science and applied technology.

About seven hundred laboratory personnel were let go during the early 1980s as a consequence of Reagan administration cost-cutting measures. However, the laboratory's multiprogram character, together with its ties through Union Carbide to the Y-12 and K-25 plants, allowed the cuts to be handled largely by transferring personnel and by not filling positions when people retired or resigned.

The first year of the Reagan revolution would prove the most unsettling for the laboratory. Deep recession in 1982 and growing federal budget deficits soon fostered less hostile views of ORNL activities within the administration. A national consensus emerged that viewed scientific and technological innovations as the nation's "ace in the hole" for breaking the cycle of rising budget deficits, high unemployment, unfavorable trade balances, and stagnant economic growth. In 1982, Herman Postma observed that support was building in government for concerted efforts to "encourage high-technology development as the best hope for the nation's economic future."

With other DOE laboratories, Oak Ridge endured the loss or retrenchment of some programs and staffing cuts of several hundred personnel yearly during the early 1980s but emerged in a stronger position later in the decade. In time, the Reagan administration abandoned efforts to dispense with DOE as well, in part because of congressional opposition, in part because of the heavy weight of bureaucratic inertia, and in part because DOE laboratories emerged as critical research centers for the Reagan-inspired Strategic Defense Initiative, or "Star Wars," program.

Thus the Reagan administration's strenuous reform efforts did not seriously sap the overall strength of the laboratory. These efforts, however, did rearrange ORNL priorities and programs. For example, Reagan policies forced the laboratory to shut down its fossil energy program and scale back its energy conservation program. When the administration terminated the government-sponsored synthetic fuel program in favor of supply-side, market-driven energy initiatives, funding for the laboratory's coal research dwindled. To maximize the return on its diminished resources, DOE decided to conduct all its coal research in laboratories formerly linked to the Bureau of Mines. Although the administration also looked unfavorably on energy conservation, the laboratory's energy conservation program survived an early round of cuts and rebounded eventually to enjoy renewed vigor.

Star Wars

In March 1983, President Reagan espoused an antimissile defense initiative that aimed to break the nuclear stalemate by shifting the battlefield to outer space, where an impenetrable defense umbrella would forever protect the United States from nuclear attack. Declaring that the strategic defense initiative would make nuclear weapons obsolete by rendering an attack futile, the president proclaimed that the proposal held promise for "changing the course of human history." Critics dubbed the initiative "Star Wars"—a flight of fancy charted by an ill-informed president that falsely promised to turn the world's fiercest technological force into its most reliable sentinel of peace.

In truth, scientific opinion was deeply divided on the long-term prospects of this proposal. Beyond the huge price tag, however, one thing was certain. Devising space satellites capable of destroying nuclear missiles would require major scientific and technological advances. Resources at DOE's national laboratories—both skilled personnel and sophisticated equipment—would be vital to any chance for success.

Managed by David Bartine, the laboratory's Star Wars research agenda, which was set by the Department of Defense, focused on three areas: reactor designs to power space satellites and particle and laser beams; flywheels for energy storage and pulsed power; and particle beams to destroy missiles from space. Studies of highly focused beams of hydrogen particles, able to destroy the electronics of a missile, evolved from the laboratory's fusion energy research, in which beams of neutral hydrogen atoms heated plasmas to high temperatures.

John Moyers headed a team from the Engineering Technology Division and other divisions for the design of a nuclear reactor to provide power bursts for the lasers and weapons aboard space vehicles. Their concept centered on the use of a boiling potassium reactor, perhaps with flywheels for energy storage. Even if never needed for national defense, the reactor might power long-distance space exploration to Mars and beyond.

Although some Star Wars research was classified, two of the laboratory's announced achievements included powerful particle beams and mirrors

Alaskan Air Command
Deployment of RL Lights

Tritium lights developed at the laboratory are installed at an Alaska airfield.

for surveillance satellites. Taking advantage of the negative-ion sources developed in fusion energy research, ORNL scientists devised the "world's highest simultaneous current density output and pulse length"—that is, a particle beam that did not spread over thousands of miles but remained tightly focused (think of a spotlight instead of a floodlight). In cooperation with scientists from K-25, Y-12, and industry, the laboratory also conducted research on beryllium mirrors and windows that would permit space satellites to sense the heat of missile launches on Earth. These mirrors and windows were devised, fabricated, and polished in Oak Ridge in cooperation with Martin Marietta Aerospace of Denver.

Star Wars research proved challenging and did not entirely cease with the end of the Cold War. The Bush administration continued to support the initiative as a key defense measure despite the new world order. Scientific proponents, such as Edward Teller, contended that Star Wars would not only protect the United States (and perhaps the world) from unprovoked terrorist attacks but could lead to yet unknown and untold applications beyond military defense. The demise of the Soviet Union, the mounting federal deficit, and the election of President Bill Clinton, however, make it unlikely that either the fervor or the funding for Star Wars will be maintained through the 1990s. Instead of being the first act in an unassailable defense initiative, Star Wars might prove to be one of the last acts of the Cold War.

As the Star Wars debate continued in the 1980s, the local media grew weary of the endless argument and eventually seemed more interested in ORNL's killer bees research than its work for the strategic defense initiative. At first glance, Star Wars and bee wars may seem to have little in common, but the efforts of researchers in both fields to track flying objects at long distances enabled them to find a common ground of scientific investigation. Newspaper journalists and television reporters enjoyed reporting ORNL efforts to detect the migration patterns of the Africanized bees, dubbed "killer bees," that moved north from Central America during the 1980s, posing a threat to national honey production.

Howard Kerr, an experienced beekeeper hobbyist, became interested in finding ways to detect and track the movements of killer bees. He and his ORNL colleagues considered tracking them with radioisotopes, spotting them with infrared devices, or identifying their presence in hives by detecting their characteristic buzzing with acoustical devices. This knowledge would provide scientists with opportunities to disrupt the bees' mating patterns. To Kerr and his colleagues, the threat that killer bees posed to honey production in North America was a serious matter; their research continued as the bees migrated across the Rio Grande into Texas during the 1990s.

Energy Systems

In 1982, the laboratory spruced itself up for the Knoxville World's Fair, building a visitor's overlook on a nearby hill and opening some facilities to tell crowds attending the fair and nearby attractions about scientific energy research taking place at Oak Ridge's national multiprogram laboratory. The laboratory also became an anchor for a proposed technology corridor championed by Tennessee Governor Lamar Alexander.

The corridor was built along Pellissippi Parkway, a highway linking west Knoxville to Oak Ridge. Its aim was to promote regional economic growth, partially through the transfer of Oak Ridge's publicly funded technology to private industries. It was hoped that Pellissippi Parkway, in time, would feature tree-lined industrial parks and glass-encased offices built to market the region's scientific and technological expertise. In effect, corridor advocates were seeking to create a Silicon Valley or Research Triangle Park in East Tennessee that would draw on the complementary skills of the region's three major institutions—Oak Ridge National Laboratory, the University of Tennessee, and the Tennessee Valley Authority.

As the World's Fair celebration began, however, the laboratory was shocked by news that Union Carbide, after nearly forty years in Oak Ridge and thirty-four years at ORNL, would withdraw as the operating contractor. Three days after the World's Fair opened in May 1982, Union Carbide management announced that the company would relinquish its contract for operating the laboratory and other Nuclear Division facilities in Oak Ridge and Paducah, Kentucky, although it agreed to serve until DOE selected a new contractor. The terse announcement read by Roger Hibbs of Union Carbide said the decision not to renew the contract resulted from the company's strategy of "concentrating its resources and management attention on commercial businesses in which it has achieved a leadership position. The corporation has no other defense-related operations."

Seventy organizations, ranging from Goodyear, Boeing, Westinghouse, Bechtel, and the University of Tennessee down to small firms, expressed initial interest in succeeding Union Carbide. After careful consideration, DOE decided to keep the Oak Ridge and Paducah facilities under a single contractor. A year after Union Carbide's decision, DOE requested proposals for operating ORNL and the other facilities, and late in 1983 it received formal responses from a half-dozen corporations and companies. It narrowed the field to three—Westinghouse, Rockwell, and Martin Marietta. In December, it accepted the proposal of Martin Marietta Energy Systems, part of the Martin Marietta Corporation known nationally for its defense and aerospace work.

Martin Marietta Corporation was formed in 1961 by the merger of Glenn Martin's aircraft company with Grover Hermann's American-Marietta Company. Aircraft pioneer Glenn Martin, a partner with Wilbur Wright, built bombers for the army during World War I; later, the firm built such famous aircraft as the *China Clippers* and *Enola Gay*. Grover

Hermann, an entrepreneur from Marietta, Ohio, had organized one of the first industrial conglomerates in the United States. Known best for its defense and aerospace contract projects, Martin Marietta Corporation managed the production of aluminum and construction materials and supervised government-sponsored defense, space, and communications initiatives.

With its corporate headquarters in Bethesda, Maryland, Martin Marietta Corporation had five operating companies employing forty thousand people at 128 sites throughout the nation. In 1984, it had major contracts for the space shuttle and MX missile designs and research laboratories located in Denver, Orlando, and Baltimore. To administer ORNL and other Oak Ridge and Paducah facilities, it formed the subsidiary Energy Systems, Incorporated.

To the relief of ORNL management and personnel, the transition from Union Carbide to Energy Systems began in January 1984 and proceeded on schedule with minimal effect on laboratory staff or activities. In April 1984, Energy Systems took full responsibility for ORNL operations along with the K-25 and Y-12 facilities in Oak Ridge and the Paducah gaseous diffusion plant in Kentucky. Later, DOE added the Portsmouth, Ohio, enrichment facilities to the Martin Marietta operations contract.

Although day-to-day operations remained much the same, the change in administration brought new long-term directions for the laboratory. Martin Marietta Energy Systems was the first contractor-operator at the laboratory without a chemical engineering background; its roots lay in prompt delivery of high-quality technology under contract with government agencies and private companies. Its agreement with DOE for operating the laboratory, moreover, contained innovative provisions, including reinvesting a percentage of its annual fee as venture capital in Oak Ridge, developing an Oak Ridge technology innovation center, and pursuing an aggressive technology transfer program.

To accelerate spin-off of Oak Ridge technology to industry, Energy Systems proposed to license DOE patent rights for technologies developed at the laboratory. In 1985, DOE approved this proposal. Energy Systems could now license to different companies the exclusive right to manufacture specific products or provide specific services based on the science and technology developed at the laboratory.

This approach would facilitate technology transfer because companies acquiring such rights could reap a profitable return without having to face direct competition. For these benefits, the companies would pay royalties or license fees to Energy Systems, which in turn would be reinvested in product refinement, prototype production, royalty shares for inventors, university programs, or other technology transfer activities. This initiative was in accord with President Reagan's policies encouraging private sector growth and economic development through the transfer of valuable scientific findings to the world of commerce.

Management Challenges

At the time of the 1984 transition, Director Postma had four associate directors administering technical activities. Donald Trauger oversaw nuclear and engineering technologies, including the Chemical Technology, Engineering Technology, Fuel Recycle, and Instrumentation and Controls divisions together with the laboratory's nuclear reactor, fuel reprocessing, safety, and waste-management programs. Murray Rosenthal supervised the laboratory's advanced energy systems programs, including the Energy and Fusion Energy divisions, along with the conservation, fossil energy, and fusion programs. Alex Zucker administered the physical sciences, including the Physics, Chemistry, Analytical Chemistry, Solid State, Engineering Physics and Mathematics, and Metals and Ceramics divisions. Chester Richmond had purview over the biomedical and environmental sciences, with the Biology, Environmental Sciences, and Health and Safety Research divisions; the Information Center complex also was assigned to him. The support and services divisions reported to the executive director, Kenneth Sommerfeld.

Biomedical and Environmental Sciences

The laboratory's biomedical and environmental sciences programs may have had the most direct influence on American life during the 1980s; at least, their environmental and health research interests dominated the national news media during the decade.

In keeping with trends at DOE, as the laboratory's energy technology focus diminished, it turned to major national environmental and health issues, and as funding for applied research increased, those divisions capable of providing it thrived. As a result, the laboratory's Environmental Sciences Division directed by Stanley Auerbach and later by David Reichle and its Health and Safety Research Division directed by Stephen Kaye flourished. By the end of the 1980s, about a quarter of the laboratory's program lay in the environmental and health fields.

The laboratory's basic ecological research continued to concentrate on understanding the processes by which contaminants moved through the environment and on identifying the ecological effects of energy production. With the National Environmental Research Park opened as an outdoor laboratory in 1980, studies of southern and Appalachian regional ecosystems continued. ORNL also expanded its hydrologic and geochemical expertise in support of DOE waste management programs to examine the effects of waste on the environment, particularly waste produced by nuclear and chemical processes.

The laboratory's study of indoor air pollution, started in 1983 by members of the Health and Safety Research Division for the Consumer Product Safety Commission, received a great deal of media attention. Laboratory surveys found that newer homes with tighter construction and improved insulation suffered indoor air pollution from substances such as formaldehyde and radon. Of special concern was radon gas, a decay product of natural uranium in the ground that seeped upward and concentrated in the more tightly sealed homes. If inhaled, it was considered a potential cause of lung cancer. Manufacturers soon were selling radon detection kits to homeowners and urging them to vent the gas from their homes if the levels of indoor radon exceeded government guidelines.

Risk assessment, whose practitioners analyzed the potential risks posed by energy technologies and industrial processes, emerged as an important field within the laboratory. Such assessment involved extensive use of computer modeling, laser optics, and advanced instrumentation to detect and examine the effects of energy- and chemical-related compounds on ecosystems. Much of this work concentrated on specific chemicals cited by EPA as potential agents of contamination.

The ecological challenges presented to the laboratory during the 1980s extended from the region and nation to the world beyond. Biomedically, long-term studies of carcinogenesis, mutagenesis, and other damages to organisms continued with major support from the National Cancer Institute and other institutes of the Department of Health and Human Services.

Within the Biology Division, research changed dramatically during the 1980s due to the advent of genetic engineering and recombinant DNA technology. Biologists learned to alter genes as simply as they had combined and separated chemicals in earlier times. This expanding capability permitted them to characterize cancer-causing genes, clarify the mechanisms for regulating gene behavior, produce scarce proteins for studies, and design new proteins. Major laboratory research initiatives included basic studies of proteins and nucleic acids, together with the mechanisms of DNA repair, DNA replication, and protein synthesis, which relate to the response of biological systems to environmental stresses.

Fred Hartman led a Biology Division group, for example, that sought to use protein engineering to improve crops. The group tried to alter plant enzymes so that they no longer used oxygen to break down carbohydrates. By improving the efficiency of enzymes, they hoped to increase plant yields for food and energy production.

As funding for basic sciences declined in favor of support for the applied sciences, the number of Biology Division researchers shrank during the 1980s to less than half the number employed during the 1960s. It retained a distinguished staff, however, and took pride in the fact that seventeen biologists who had worked at the laboratory were elected to the National Academy of Sciences.

The laboratory's emphasis on the production, development, and use of radiopharmaceuticals contributed to improved public health in several ways during the 1980s. F. F. Knapp's nuclear medicine group in the Health and Safety Research Division made news by developing new radioactive imaging agents for medical scanning diagnosis of heart disease, adrenal disorders, strokes, and brain tumors. Stable isotopes, produced in calutrons in the Chemical Technology Division, were converted into radioisotopes such as thallium to provide the tracing material for millions of heart scans. By the end of the 1980s, DOE estimated that nearly 100 million Americans annually received improved diagnosis or treatment partly as a result of medical isotope research and production at the laboratory and other DOE facilities.

Another medical advance arose from work at the Solid State Division's Surface Modification and Characterization Collaborative Research Center. Here, various ion-beam and pulsed-laser techniques were used to improve and characterize the properties of materials, giving them harder surfaces, more resistance to wear and corrosion, and improved electrical or optical properties. Applied initially to such semiconducting materials as silicon for solar cells, these techniques later proved beneficial in the development of other new materials including surgical alloys.

Since the 1950s, thousands of patients had been fitted with artificial hip joints made of a titanium alloy. Body fluids, however, caused corrosion and wear that required replacement of the devices after about ten years. At ORNL during the 1980s, James Williams and collaborators implanted nitrogen ions into the titanium alloy to modify the surface. This made the artificial joints more resistant to the wear and corrosive action of body fluids, significantly increasing the resilience and lifetime of such joints. This process was incorporated into a new medical products line marketed by Johnson and Johnson Corporation. Star athlete Bo Jackson became the most famous beneficiary of the improved artificial hip joint.

New devices in the Biology and Health and Safety Research divisions made possible the imaging of single atoms and of DNA strands during the 1980s. Scanning tunneling microscopes, developed in 1980 and first used for research on semiconductor surfaces, were built at the laboratory during the decade. These microscopes, which gave new meaning to the word microscopic, could image supercoiled DNA molecules, showing structural changes and the binding of proteins and substances to the genetic strands. Operated by David Allison, Bruce Warmack, and Thomas Ferrell, the new electron and photon microscopes promised to assist in mapping and determining the sequences of genes in DNA as part of the Human Genome Project, thus opening new frontiers in biological research.

A team of Environmental Sciences and Chemical Technology researchers sought to use microorganisms in bioreactors to rid the environment of polychlorinated biphenyl (PCBs) and other toxic wastes.

Experiments along Bear Creek in Oak Ridge indicated that aerating and watering PCB-contaminated soil encouraged the growth of microorganisms that could digest PCBs and turn them into less toxic substances. This success led to additional investigations into bacterial capabilities for digesting and converting other toxic materials.

For many years, researchers in the Health and Safety Research Division analyzed the accuracy of personnel dosimeters for the laboratory and outside agencies. Other agencies mailed dosimeters to the laboratory, and the devices were checked by exposing them to measured radiation at the health physics research reactor. In 1989, the laboratory opened a radiation calibration laboratory for checking dosimeters and the accuracy of radiobiological data. This new facility helped fill the research needs stymied by closure of the health physics research reactor.

Advanced Energy Systems

Murray Rosenthal's advanced energy systems, including the fossil energy, conservation, and fusion programs, suffered loss of program support during the early Reagan years. Relying on supply-side economics and market forces to meet energy demands, the Reagan administration dispensed with most of the fossil energy program, severely curbing fossil energy research at the laboratory.

As for energy conservation, so popular during the Carter administration, one official of the Reagan administration declared that it simply meant "being too hot in the summer and too cold in the winter" and contended that increasing energy prices would provide the only incentive needed for conservation. The administration initially mandated major cuts in conservation research funding, forcing the abrupt termination of some energy conservation projects at the laboratory. Congress, however, restored some of the budget reductions, and the laboratory's conservation program flourished again during Reagan's second term.

Studies of improved building insulation for energy conservation continued, and George Courville, Michael Kuliasha, and William Fulkerson sought the creation of a Roof Research Center in the Energy Division. A cooperative effort by DOE and the building industry, this center opened in 1985 to measure heat transfer through roofing structures, to assess how these structures reflected or absorbed solar energy, and to project how long the insulating material would last. In climate-simulation facilities, composite roof segments were instrumented and tested, providing data for the computer modeling of roofing designs. Directed by Paul Shipp and Jeff Christian, the roofing research identified significant convective heat losses in common blown attic insulation and worked with the building insulation industry to devise more efficient systems.

In cooperation with the National Bureau of Standards and industry, ORNL studies of improved home appliances produced significant results as well, notably in the development of absorption heat pumps for heating and cooling that could be powered with natural gas instead of electricity. The Energy Division's Michael Kuliasha and Robert DeVault managed subcontracts with industrial firms to improve and commercialize these heat pumps. Thanks to these and other innovative ventures, the laboratory's conservation and renewable energy program recovered its losses; in fact, its annual budget rose from $28 million at the start of the decade to $46 million by 1988.

Physical Sciences

The laboratory's physical science research efforts, under the direction of Alex Zucker and later William Appleton, focused on nuclear physics, chemistry, and materials science. To conduct their experiments, researchers turned to the Holifield Heavy Ion Research Facility, neutron-scattering facilities at the high-flux isotope reactor, and the Surface Modification and Characterization Collaborative Research Center. Zucker also hoped his staff would become instrumental in pushing the boundaries of material sciences through a new High Temperature Materials Laboratory housing modern equipment for testing the properties of materials needed in high-temperature applications such as highly efficient engines. The keynote of this program was the development of existing and new user facilities in partnership with industry and universities.

Basic research on the chemistry of coal and solvent extraction continued at the laboratory, but the loss of most of the fossil energy program took several divisions into the field of bioconversion as a poten-

Charles Scott and Charles Hancher at a fluidized-bed bioreactor used to reduce nitrate concentrations in water.

tial source of energy and improved waste disposal management. Bioconversion research sought to use microorganisms to convert organic materials—sewage, solid wastes, woody biomass, coal, or corn—into fuels. Rather than liquefying coal with heat and pressure, for example, Charles Scott and teams in the Chemical Technology Division turned to bioreactors in which microorganisms convert coal to liquids. In another case, the laboratory cooperated with the A. E. Staley Corporation, a corn products company with a plant near Loudon, Tennessee, to improve the fermentation of corn using a fluidized-bed bioreactor in which bacteria converted almost all the sugar in corn into ethanol, which can be used as a petroleum substitute.

Materials research rose to the forefront of the laboratory's efforts in physical sciences during the 1970s and 1980s. The laboratory was a pioneer in alloy development, high-temperature materials, surface-modification technology, specialized ceramics, and the development of composite materials. These successes placed it in a position to contribute directly to industrial technology applications. Welding science can serve to illustrate.

In the nuclear power industry, proper welding is as critical to safety as it is in most other industries—perhaps even more so. The welding and brazing group established at the laboratory in 1950, therefore, had many opportunities to improve welding technology, and it gained world-wide recognition for its contributions.

National energy production has been hampered when poor welds shut down nuclear power plants, coal-fired plants, and petroleum refineries. In 1985, when Alex Zucker asked welding specialist Stan David and physicist Lynn Boatner to review ORNL research on composite materials, they concluded that a multidisciplinary attack on fundamental welding problems could be fruitful. Faulty welding continues to undermine the reliability and safety of nuclear reactors in particular. As a result, electric utilities have leaned on ORNL researchers to test and improve the performance of welds, which have proven critical to the future of the nuclear industry.

The user facility attracting the greatest attention during the 1980s was the High Temperature Materials Laboratory. First proposed in 1977, it required a decade of efforts by Fred Young, John Cathcart, Victor Tennery, James Weir, James Stiegler, Carl McHargue, Ted Lundy, and associates to get the twenty-million-dollar user facility completed. Deferred by the Reagan administration in 1981, persistent academic and industrial interest overcame the administration's initial resistance and abruptly shifted the project to the front burner. Funded in 1983 by DOE's Energy Conservation Program, it opened in April 1987 and housed forty-nine laboratories and seventy-two offices for staff and visitors. Just as the underground building for the Joint Institute for Heavy-Ion Research, completed in 1979, seemed a monument to the brooding Carter administration, the sleek, wing-tipped architectural design of the High Temperature Materials Laboratory seemed emblematic of the Reagan administration's unbounded optimism.

The High Temperature Materials Laboratory fostered exactly the sort of scientific research the Reagan administration demanded. Its modern instruments, microscopes, and furnaces made possible advanced ceramics research designed to increase the competitiveness of the United States in international markets. Advanced engines may operate at such high temperatures that ordinary metal alloys melt, but stronger, heat-resistant ceramic or intermetallic components can beat the heat and keep on clicking. Research at the High Temperature Materials Laboratory thus promised to improve vehicle, aircraft, and rocket engines for maximum fuel efficiency. These materials also could promote development of superconducting ceramic magnets, advanced electronic components, and lightweight armor for tanks and other military applications.

When President Reagan visited the University of Tennessee in Knoxville in 1985, Director Herman Postma had an opportunity to tell him about ORNL activities. Using its development of sturdier artificial hip joints as an example, Postma emphasized the laboratory's new role as a user facility seeking to expand partnerships with universities and industry. Instead of closeting its research behind a fence, the laboratory had become a place that opened its doors to innovation and collaboration. "We have large and unique facilities in Oak Ridge, and we open them to users from throughout the country," he told the president. "We have also helped the University of Tennessee to establish centers of its own that are privately funded by industry," he went on. "Perhaps most importantly, we share accomplishments."

The laboratory's responsiveness to a new set of national needs brought it out of the doldrums of the early 1980s into renewed prosperity. After setbacks during Reagan's first term, the laboratory's overall operating budget rose to $392 million in 1988, slightly larger in constant dollars than it had been in 1980.

Seed Money Spreads

Postma viewed the seed money program for unfunded exploratory studies an undiluted success. Since the program's beginnings in 1974, seed money projects had brought about four dollars in new research funding to the laboratory for every dollar invested.

To build on this success, the laboratory in 1984 established two new exploratory research funding opportunities: a Director's Research and Development Fund for larger projects and a Technology Transfer Fund to encourage commercially promising research. It is our "strong view," Postma asserted, "that the best judges of technical opportunities are those doing the work and their peers."

Seed money projects provided grants of up to one hundred thousand dollars for one year's work, long enough for the work to produce results that could acquire attention and funding from a sponsor. The Director's Research and Development Fund, created in 1984, supported larger projects, ranging from one hundred thousand to six hundred thousand dollars, selected from proposals submitted by laboratory divisions.

Among early projects supported by the Director's Fund was a project managed by Donald Trauger and James White to assess promising smaller, safer nuclear reactors to determine whether they could be commercially developed. Reactor designs under study included liquid-metal–cooled reactors, process-inherent-ultimately-safe (PIUS) reactors, small boiling-water reactors, and high-temperature, gas-cooled, and pebble-bed–fueled reactors.

Robotics

Another Director's Fund project of 1984 was CESAR, the Center for Engineering Systems Advanced Research. Headed by Charles Weisbin, this center focused on computer problem solving through artificial intelligence resembling human reasoning. The "reasoning" generated by machine-produced artificial intelligence was to be exercised through remotely controlled robots capable of working in such hostile environments as outer space, war zones, areas contaminated by radiation, or coal mines.

Since the days when the laboratory recovered plutonium from the graphite reactor and Waldo Cohn initiated radioisotopes production, remote control of operations in hostile environments had been an ORNL specialty. Elaborate servomanipulators had been designed and built to accomplish work from behind the protection afforded by concrete or lead walls. Moreover, Mel Feldman, William Burch, and leaders of the Fuel Recycle Division had become interested in using robots to accomplish nuclear fuel reprocessing through teleoperations from a distance, or, as Feldman put it, "to project human capabilities into hostile workplaces without the actual presence of humans."

In the mid-1980s, the laboratory formed a telerobotic task force, under Sam Meacham, to acquire new programs and sponsors for re-

An experimental robot
built at the laboratory
delivers the 1985 state
of the laboratory
address to Director
Herman Postma.

search in robotics and teleoperations. For this effort, the laboratory received support from NASA to develop the man-equivalent telerobot for satellite refueling and space-station construction. The laboratory also received funding from DOE's new Office of Civil Radioactive Waste Management to assess the applications of robotics and remote technology for the proposed monitored retrievable storage facility that was intended to provide temporary storage for high-level nuclear waste.

Members of the Fuel Recycle, Instrumentation and Controls, and Engineering divisions contributed to the robotics program. Also, the Engineering Physics and Mathematics Division broadened its technological bases in robotics and artificial intelligence. These initiatives led to the Robotics and Automation Council, the precursor of the laboratory's Robotics and Intelligent Systems Program headed first by Charles Weisbin and then by Joseph Herndon.

In 1985, the laboratory began tests of a motor-driven robot that could sense its surroundings through sonar and machine vision and respond to computer commands relayed by radio. Investigators Reinhold Mann and William Hamel improved the basic design to create one of the world's most computationally powerful robots. The size of a small car, it could sense its surroundings, deal with unexpected events, and learn from experience.

Acquiring funding from DOE, NASA, the army, and the air force for robotics research, the laboratory formed the Robotics and Process Systems Division in the early 1990s and initiated research aimed at devising remotely controlled robots with "common sense." One early accomplishment was the robotic mapping of waste-filled silos at DOE's Fernald, Ohio, facilities. The robotic effort helped DOE conduct its complex remediation work more safely and cheaply.

In the words of one laboratory scientist, robotics research resembled a "Buck Rogers adventure." For children of today's generation, Captain Nemo, not Buck Rogers, may be a more apt analogy from the world of entertainment. But for both young and old, the effort again proved science's unique ability to enliven the imagination by turning the fantastic into reality.

Chernobyl's Fallout

Oak Ridge, the rest of the United States, and the entire world watched and worried in April 1986 as a radioactive cloud from the massive reactor failure at Chernobyl in the Soviet Union circled the globe. The Three Mile Island accident in Pennsylvania had taken place seven years before but remained a fresh memory for many people concerned about the safety of nuclear power.

The more serious accident at Chernobyl renewed public fears and further dampened hope of reviving commercial nuclear power in the United States. The Soviet tragedy also caused a massive DOE reexamination of reactor safety throughout the nation, including detailed inspection of reactors at the agency's nuclear facilities. An industry that had been reeling from mistakes and mishaps for two decades now went into a tailspin.

DOE funding for nuclear power research at the laboratory had been severely curtailed during the 1980s, even before the Chernobyl accident. "ORNL used to be thought of as a nuclear energy laboratory, a facility whose main mission was fission," Postma remarked in 1986. "That obviously is not the case now." The laboratory's reactor research budget plummeted from $50 million in 1980 to $13 million in 1986, representing only 3 percent of the laboratory's total budget.

Laboratory personnel monitor tests of the high-flux isotope reactor's pressure vessel in 1987.

A few weeks after Chernobyl, Postma appointed a committee chaired by Donald Trauger to review safety at the aging high-flux isotope reactor. After locating and assessing the data, the committee learned that twenty years of neutron bombardment had embrittled the reactor's vessel more than had been predicted. In November 1986, the laboratory shut down the reactor to conduct a thorough investigation. These precautionary steps had severe effects on the laboratory's research agenda: They delayed neutron-scattering research and neutron activation analysis, slowed irradiation testing of Japanese fusion reactor materials, and reduced radioisotope production for medical research. The loss of californium-252 production, an isotope vital for cancer research and treatment and for industrial uses, proved especially critical.

Increasingly concerned about reactor safety management, DOE shut down all reactors at the laboratory in March 1987. To oversee a safe restarting of at least some of the reactors, Fred Mynatt became the associate director for reactor systems, and responsibility for reactor operations was assigned to a new Research Reactors Division. For the first time since its inception in 1943, however, the laboratory in 1987 had no nuclear reactors in operation.

From Arsenal to Engine

Although no longer strictly a nuclear laboratory, the multiprogram laboratory at Oak Ridge during the 1980s savored the essence it had inherited. "The essence of a laboratory is that it experiments," Postma said, "it explores, it hurls itself against the limits of knowledge. In short, it tries. Often it fails."

Still, the change in national administrations in 1981 and the switch of contractor-operators in 1984 sparked a new phase of research within the laboratory. The cornerstone of this new age of accomplishment was the expanding partnerships with industries and universities. Between 1980 and 1988, the list of official DOE user facilities at the laboratory increased from three to twelve and the number of guest researchers tripled. In 1991, the laboratory had 3,600 guest researchers at work in its user facilities; 30 percent of these guests came from industry, compared with 5 percent in 1980.

Technology transfer became the second highlight of the laboratory's surprising renaissance during the Reagan and then the Bush administration. By transferring the laboratory's scientific and technological advances speedily into the private sector, the administration and Martin Marietta Energy Systems hoped to boost the national economy and improve the competitiveness of U.S. products in international markets. As President George Bush summed it up during a 1992 visit to Oak Ridge, the multiprogram laboratory was being transformed from "the arsenal of democracy into the engine of economic growth."

As the Cold War fades into history and international economic competitiveness becomes the hallmark of a nation's prowess, the laboratory's ability to negotiate, in the words of former President Bush, the challenging transformation from "military arsenal" to "economic engine" is likely to determine how well it serves the nation's interest in the twenty-first century and beyond. This challenge of transformation was one of few certainties that laboratory officials could count on as they approached an unpredictable future that would no longer be shadowed by the Soviet menace.

Chapter 9

Global Outreach

As the laboratory approached its fiftieth anniversary, science—always an international enterprise—assumed even broader global dimensions. Just as national borders were drifting away in the business world, the interests of basic and applied scientists transcended national boundaries and entered the global arena at an unprecedented scale and scope. Events at the laboratory during the 1980s and early 1990s reflected this global transition.

The U.S. Agency for International Development, for example, called on the laboratory to help Third World countries satisfy their growing appetites for energy in the face of persistent fuel and resource shortages. Improved energy efficiency is a key to their quest, and ORNL researchers have sought to provide the Third World with valuable technical assistance based on decades of energy research and development here at home.

In its quest for abundant fusion energy, ORNL intensified its scientific cooperation with laboratories in other nations in an effort to get a grip on fusion energy's elusive brew. Its environmental research, which focused originally on radioecology, expanded to encompass worldwide environmental threats. And the laboratory's life sciences divisions united with others in an international initiative to map and sequence the human genome. In short, having started in 1943 as a national scientific laboratory devoted to nuclear research and development, by 1993 ORNL had evolved into a global science center.

As its global missions proliferated, the laboratory's top management experienced a transition that reflected ORNL's desire to keep pace with these new research frontiers. George Bush, who became president in 1989, had spent most of his career as a federal employee. Unlike Reagan's (and even Carter's), his administration did not place unbridled opposition to the federal government at the center of its domestic agenda. In fact, when beyond the hot lights and intense scrutiny of the media, Bush often filed his campaign rhetoric and proposed to use government agencies, including DOE laboratories, to advance his goals.

Specifically, Bush augmented the duties of the presidential science advisor. The person he chose for this position, D. Allen Bromley, assumed the lofty title of the assistant to the president for science and technology and the director of the Office of Science and Technology Policy. More importantly, Bromley had ready access to the president, which gave him the clout to rejuvenate many existing committees that had ceased to function effectively—notably the Federal Coordinating Council for Science, Engineering, and Technology and the President's Council of Advisors for Science and Technology.

The Bush administration also devised a new overall strategy for federal scientific research and development, which it placed under the banner of "presidential initiatives." When such initiatives were announced in global climate modeling, high-performance computing, advanced materials and processing, and biotechnology, the laboratory responded with proposals and programs.

To head DOE, Bush selected Admiral James Watkins, a veteran of Rickover's nuclear navy. Watkins, in fact, had attended the Oak Ridge reactor school during the 1950s and later recalled that "it was the bright minds of the academics at Oak Ridge, not the blue suit people, who inspired me to go forward in the Navy." From nuclear submarine and ship commander, he rose to chief of operations before retiring from the navy to become secretary of energy.

This national transition in energy-related policies and positions was paralleled by changes in ORNL management. After fourteen years at the helm, Herman Postma transferred to the executive ranks of Martin Marietta Energy Systems in early 1988. While Associate Director Murray Rosenthal chaired a committee to recommend Postma's successor, Alex Zucker acted as laboratory director throughout 1988. A high-energy physicist, Zucker had come from Yale University to the laboratory in 1950 to launch its cyclotron program. A naturalized citizen born in Yugoslavia, he possessed an international viewpoint that inspired closer association with the global scientific community.

Although not troubled by severe budgetary constraints such as those that plagued the early 1980s, Zucker inherited several "crises" demanding ORNL attention. The least troublesome crisis focused on fears that international terrorism might extend into the United States, even through the gates of Oak Ridge and into the laboratory's research facilities. Charles Kuykendall, ORNL Protection Division director since 1979, marshaled his division's resources to protect the laboratory against potential terrorist assaults, adding an emergency preparedness department and opening a center for high-technology security. Although the laboratory was never remotely threatened by international terrorism, the new safeguards proved useful, especially when the 1991 Gulf War heightened concerns about terrorism and when President Bush visited the laboratory in 1992.

A second and longer-lived crisis of the late 1980s involved environmental safety and health in DOE facilities. Under new, more stringent laws and regulations, often spurred by agency neglect of such issues in the past, federal and state environmental officials monitored both remedial and preventive measures designed to protect human health and the environment on the Oak Ridge reservation and in the surrounding communities and counties. At the laboratory, scores of air and groundwater monitoring devices were added and dozens of environmental safety specialists were hired to comply with the stricter standards. As part of this initiative, the laboratory also investigated and tested new methods of waste management. While pursuing these initiatives—and under strong pressure for public disclosure—a legacy of environmental neglect was uncovered at the laboratory.

Estimates indicated that environmental restoration costs at the laboratory could reach $1.5 billion, and that restoration costs at all DOE installations could exceed $300 billion and take more than thirty years to complete. The laboratory's long-standing leadership in environmental restoration technology, it was hoped, could partially offset these staggering costs, provide the laboratory with new areas of research, and counterbalance funding cutbacks in other ORNL research fields, most notably in nuclear energy, space, and weapons developments. Officials even suggested that Oak Ridge might become an international center of excellence in waste management.

Prospects of turning this crisis into an opportunity, however, do not minimize the enormous sums of money and the army of personnel that will be applied to the cleanup campaign, which may prove in the years ahead to be as encompassing and complicated as the Manhattan Project was in years past. The Cold War is over, but its impact on the domestic front will be felt for many decades.

A third crisis afflicting the laboratory in 1988 involved safety assurance for its nuclear reactors. In the aftermath of worldwide concerns for the safety of nuclear power sparked by the Chernobyl disaster, DOE had closed the laboratory's five reactors in 1987 for comprehensive safety reviews. The Oak Ridge research reactor had been scheduled for decommissioning, so its loss was not significant. But laboratory officials thought it imperative that the high-flux isotope and tower shielding reactors be reactivated quickly to alleviate radioisotope shortages and permit the resumption of scientific experiments. Officials also identified important ORNL research programs that depended on the health physics and bulk shielding reactors, but the costs of the prescribed environmental, safety, and health improvements precluded their future operation.

Pressed by DOE, Zucker initiated a campaign to improve quality assurance. The laboratory's Quality Department (formerly Inspection Engineering) increased its work force to twenty-eight professionals. This staff helped clear the way for the restart of the high-flux isotope and tower shielding reactors, prepared quality assurance documentation in accord with new standards, and sought to correct deficiencies identified by internal and external quality assurance audits by DOE, Energy Systems, and other agencies.

Regaining public trust and confidence in reactor safety has proven a slow, tedious, and fragile process. Citizen support for nuclear research and development, in fact, will never return to the levels of enthusiasm expressed in the 1940s, 1950s, and 1960s. Despite this erosion of public support, by the early 1990s the laboratory's restarted reactors enabled Oak Ridge to continue on its research journey into the mysteries of the atom—a journey that is now more than half a century old.

On a less controversial note, during Zucker's year at the helm the laboratory boosted its position as an international leader in materials research. By integrating applied materials research (lodged chiefly in the Metals and Ceramics Division) with basic research (found mostly in the Solid State and Chemistry divisions), the laboratory placed the entire spectrum of materials research—from developing new ceramics and alloys to operating new furnaces—under one administrative roof. Under this new administrative umbrella, the laboratory hoped to achieve a broader understanding of surface phenomena and physical properties. Such knowledge, in turn, could be applied in a thousand ways— from improving the efficiency of electricity transmission to enhancing the speed and safety of ground transportation.

In addition to coping with the challenges facing the laboratory in 1988, Zucker concentrated on reassuring the staff that advancing science and technology would remain the laboratory's principal goal. In the late 1980s, ORNL scientists had expressed increasing concerns that the laboratory's preoccupation with environment, health, and safety, coupled with the intense consideration given to compliance in setting contractor-operator award fees, would compromise ORNL research and make it a less-than-central consideration. Unlike benchmarks set for compliance, laboratory researchers contended, cutting-edge science is a high-risk proposition whose progress is difficult to assess every six or twelve months.

To alleviate this concern, while at the same satisfying the requirements of the awards system, the laboratory sought to emphasize its user facilities and opportunities in technology transfer. Nevertheless, nurturing the freedom that scientific research demands within a system that relishes concrete timetables and goals remains a difficult task.

By the time Alvin Trivelpiece became the new laboratory director in early 1989, the laboratory had improved its emergency response system, promoted some innovative waste-management technologies, and stood ready to resume reactor operations. There would be no quick fix, however, to the waste-management and reactor operations crises, both of which would help define the laboratory's agenda in the 1990s. In fact, a substantial portion of the laboratory's research and experimentation through the first half of the decade would be devoted to fixing past mistakes, rather than launching new research initiatives.

Director Alvin Trivelpiece.

Trivelpiece Reorganizes

In his first address as director in 1989, Alvin Trivelpiece outlined the broad themes of his administration. "As a national laboratory, we need to be able to respond both to inflicted change and to the changes we may cause to occur," he declared. "We need to be a competitor," he added, "we need to be serious about competing, and to be taken seriously as a competitor in the world's research and development efforts."

Preparing to meet these challenges, Trivelpiece reorganized ORNL management. Zucker was appointed associate director for nuclear technologies, a post he held until moving to the Energy Systems executive staff in 1992. James Stiegler replaced him, and his "directorate" was renamed Engineering and Manufacturing Technologies. Murray Rosenthal was named deputy director for administration and was given primary responsibility for health, safety, and the environment. William Fulkerson succeeded Rosenthal as associate director for advanced energy systems,

Chester Richmond continued as associate director for biomedical and environmental sciences, and William Appleton was designated associate director for physical sciences and advanced materials.

As part of the reorganization, Trivelpiece supported several program initiatives and organizational changes to foster new laboratory missions and directions. He breathed new vitality into the advanced neutron source (ANS) project, which the laboratory hoped would lead to the construction of its first new research reactor in more than twenty-five years. For example, he divided ANS project responsibilities into reactor operations and scientific research, corresponding to the two major challenges ORNL staff faced in justifying federal expenditures: How reliable would the reactor be, and what kind of research would it support? With Colin West as project director and John Hayter as scientific director, the advanced neutron source became a top laboratory priority.

A strong proponent of the now defunct superconducting supercollider, Trivelpiece also encouraged vigorous laboratory participation in that project's design and development, largely through creation of an Oak Ridge Detector Center with Tony Gabriel as its director. Acknowledging worldwide scientific concern for the potential impact of global warming, Trivelpiece also encouraged creation of a Center for Global Studies under the direction of Robert Van Hook and Michael Farrell.

The new director also strengthened the Office of Planning and Management under Truman Anderson. To meet the needs of the increasing number of outside guest scientists and users and to coordinate the Cooperative Research and Development Agreements (CRADAs), an Office of Guest and User Interactions was established.

Another Trivelpiece initiative enhanced scientific computing at the laboratory by establishing an Office of Laboratory Computing under Carl Edward Oliver. Citing the expertise developed in parallel computing in the Engineering Physics and Mathematics Division, DOE designated ORNL as a High Performance Computer Center—one of only two laboratories granted this responsibility. To promote the use of high-performance computers, a new Center for Computational Sciences also was established.

In partnership with universities and other laboratories, these supercomputers, it was hoped, would help Oak Ridge confront key scientific challenges of the late twentieth century—for example, the unknown

The laboratory's Computer Science Research Facility opened in 1991.

frontiers in global climate research, human genome sequencing, high-energy heavy-ion physics, and groundwater transport of contaminants, all of which required not only new research paradigms but the ability to process and digest mountains of data.

Trivelpiece also enlisted the laboratory in a campaign spearheaded by Secretary Watkins and President Bush to foster science and mathematics education. In February 1990, he appointed Chester Richmond director of the laboratory's science and math education programs, an announcement that coincided with President Bush's visit to Knoxville to boost public support for science education. Under this initiative, the laboratory expanded its educational programs designed to promote elementary and secondary science education largely through hosting student workshops and teacher-training seminars. In addition, the science education program strengthened ORNL cooperation with minority educational institutions in an effort to attract new students into the world of science. More than sixteen thousand precollege students visited the laboratory in 1991, many participating in weekend academies for computing and mathematics.

When Richmond moved to the science and math education programs in 1990, David Reichle succeeded him as associate director for biomedical and environmental research, later expanded to include the Energy Division and renamed the Environmental, Life, and Social Sciences Directorate. By 1992, these "soft" sciences had experienced significant growth, which enabled the laboratory to immerse itself in research on global environmental change, economic competitiveness, and human health.

Reactor Management

Restarting its reactors topped the laboratory's agenda in the late 1980s. After extensive safety investigations, DOE's Oak Ridge operations manager, Joe La Grone, recommended reactivating the high-flux isotope reactor in late 1988. And, in March 1989, Admiral Watkins surprised a Senate committee by concurring with La Grone's assessment and announcing his decision to partially resume reactor operations at Oak Ridge under strict guidelines and oversight.

Managed initially by Robert Montross and later by Jackson Richard, the high-flux isotope reactor was restarted in April 1989. After experiencing initial operational difficulties, the reactor ran smoothly at 85 percent of its original power. The laboratory also restarted its tower shielding reactor in December 1989, allowing shielding studies for breeder reactors funded by DOE and Japan to proceed. The laboratory, however, mothballed its bulk shielding, health physics, and Oak Ridge research reactors, and initiated steps to decommission them, although many ORNL researchers believed the health physics reactor remained a national asset that deserved a better fate.

Age of Materials

In 1989, the National Research Council published a comprehensive study of materials science and engineering subtitled *The Age of Materials*. The study offered a detailed assessment of the critical roles materials science and engineering would play in the future competitiveness and

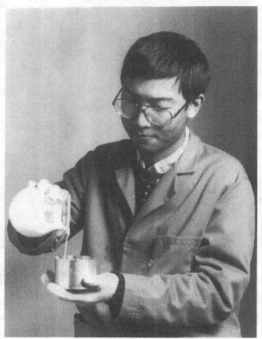

Albert Young pours an aluminum powder and pre-gel solution into a mold, the first step in the laboratory's gel-casting process for ceramic production.

prosperity of the United States. In the report, the laboratory's systematic and multidisciplinary materials science program served as a case study of why materials were technologically and economically important, and why the 1990s seemed destined to become the "age of materials."

Materials science, which had begun in earnest during the laboratory's nuclear airplane project in the 1950s, had slowly evolved from a program defined by disparate agendas into a cohesive and comprehensive research initiative. The Solid State Division, launched in 1950, initially examined radiation effects on existing materials but expanded over the decades to explore potential uses of new materials. The Metals and Ceramics Division, begun in 1948, steadily moved into broad research and development efforts that focused on advanced alloys and ceramics.

The work of these two divisions yielded a fruitful marriage of research and application in a multidisciplinary setting. Fundamental research provided the underpinnings for materials characterization and analysis. Experimentation led to the development of alloys able to withstand thermal stress, pressure, and long-term radiation. These alloys, in turn, soon found valuable commercial applications. Ion-beam facilities, for example, built to simulate reactor damage in materials, could be used to fabricate semiconductor devices and solar cells.

ORNL staff also contributed to NRC assessment of materials science. William Appleton, for example, chaired the council's solid-state sciences committee, and James Stiegler cochaired an assessment panel. In brief, the laboratory surfaced as an active participant in the emerging "age of materials."

Global Environmental Challenges

ORNL efforts to quantify and resolve threats to the global environment began as early as 1968 when Jerry Olson of Environmental Sciences Division initiated studies of carbon dioxide levels in the world's atmosphere. In 1976, Alex Zucker expressed concern about global warming—that is, the potential for temperatures to rise largely because of increased carbon dioxide levels in the Earth's atmosphere—and he assembled a team composed of Olson, Ralph Rotty, Charles Baes, and Hal Goeller, to study the problem and recommend appropriate ORNL actions.

Observing that carbon dioxide concentrations in the air had increased steadily since the beginning of the Industrial Revolution, the team identified the sources and sinks (or repositories) of carbon dioxide, pinpointing the crucial role of oceans in absorbing carbon dioxide from the atmosphere and demonstrating the great uncertainties connected with the problem.

With DOE support, the laboratory began to analyze the emerging global environmental concerns related to energy use. The burning of fossil fuels and clearing of forests were cited as the prime causes of the steady buildup of carbon dioxide in the atmosphere. Fossil fuel burning also was linked to the formation of acids (sulfur and nitrogen oxides) in the atmosphere, which rain down on forests hundreds of miles from their diverse sources.

During the late 1970s, Henry Shugart and David Reichle proposed to DOE a study of the global carbon cycle and its relationship to fossil fuel burning. This proposal was one of several that ultimately led DOE to launch a major global carbon dioxide program. With Reichle, John Trabalka, and Michael Farrell of the Environmental Sciences Division providing leadership, the laboratory adopted an interdisciplinary research strategy to identify the sources, migration, distribution, and consequences of global warming and acidic rain deposition. This effort, in turn, sparked vigorous experimentation at the laboratory on global biogeochemistry.

Laboratory scientists used computer modeling to estimate how additional accumulations of carbon dioxide in the atmosphere might induce future global climatic changes. Some models predicted intense global warming, with potentially calamitous effects on trees and crops. In the field, ORNL scientists examined tree rings and fossil pollen grains taken from lake sediments to detect past climatic conditions and trends. For example, using radiocarbon-dated sediment taken from Tennessee ponds, Hazel Delcourt and Allen Solomon reconstructed changes in regional vegetation over sixteen thousand years. With this paleoecological evidence, they projected the future effects of carbon dioxide concentrations on vegetation and the climate.

The greenhouse effect and acid rain were truly global challenges, and quantifying their results and devising potential solutions required an understanding of complex physical, chemical, and biological processes on a global scale. The laboratory's approach, therefore, expanded to include global monitoring, measurement, and modeling using the largest, fastest computers available. The laboratory took the lead in formulating global carbon simulation models and became responsible for managing DOE research efforts, subcontracting studies to universities and other laboratories and establishing the Carbon Dioxide Information and Analysis Center to compile and disseminate data.

To investigate acid rain and its effects, the Environmental Sciences Division installed rainmaker simulator chambers in a greenhouse and programmed them to control raindrop size, intensity, and chemical composition; for comparison purposes, they built an identical system using unpolluted water. These experiments examined the consequences of prolonged ecosystem exposure to precipitation polluted by sulfur and nitrogen oxides, ozone, and other materials. The accumulated data helped set regulatory standards for environmental protection under the 1990 amendments to the Clean Air Act.

David Shriner examines plants in a rainfall simulator at the laboratory to assess the effects of acid rain.

In the late 1980s, the Electric Power Research Institute and other agencies funded ORNL studies that examined the effects of acids on streams in the Appalachian, Great Smoky, and Adirondack mountains. Managed by Ernest Bondietti, this project sought the cooperation of a dozen universities in the eastern forest region. Early results indicated that acids in mountain streams had natural geologic sources in addition to human-induced sources. In short, the studies indicated that solving the acid rain problem would entail more than stemming the flow of industrial pollution.

At the laboratory's Walker Branch Watershed, Dale Johnson and Daniel Richter conducted forest-nutrient cycling research on the soil-leaching effects of acid deposition. In 1992, the laboratory announced the watershed would be the site of the first large-scale field studies of the effects of global warming on forest growth.

This and other research supported steady growth in the laboratory's environmental sciences program. With about two hundred full-time employees and more visiting university faculty and students than any other division, the Environmental Sciences Division nurtured a world-class reputation rooted in its investigations of international environmental issues.

In July 1989, Trivelpiece announced formation of a Center for Global Environmental Studies to be managed by Robert Van Hook and Michael Farrell from the Environmental Sciences Division. "Its goal," Trivelpiece said, "would be to achieve better understanding of global air, land, and water environments and more accurately predict the consequences of human activities on the world's ecological balance." The center would concentrate on the causes and effects of such global challenges as greenhouse warming, ozone depletion, acid rain, and deforestation.

By the early 1990s, the laboratory had conducted major studies on the perplexing problem of ozone depletion, or what the media commonly called the "ozone hole." In cooperation with industry, the laboratory joined in the search for acceptable substitutes for the suspected chemical culprit, chlorofluorocarbons (CFCs), used widely in refrigerants, insulation, and commercial solvents. Studies at the laboratory's Roof Research Center in the Energy Division, for example, focused on testing foam-board insulation made with ozone-safe CFC substitutes.

Hot and Cold Con-fusion

Fusion energy researchers were shocked when two chemists from the University of Utah announced in a March 1989 press conference that they had achieved cold fusion, or fusion at room temperature. By passing electricity through chunks of palladium metal immersed in jars filled with electrically charged heavy water, they said that they had produced a fusion reaction. If true, the discovery offered an inexpensive alternative to "hot" fusion as an unlimited energy source.

Trivelpiece learned of this startling announcement on the front pages of his weekend newspaper—an unlikely source for a scientist to discover a research breakthrough that challenged the canons of his profession. "I used the only scientific tool available to me that weekend—a push-button telephone," he later remembered, "and called everyone I knew who might be able to help me and I tried to find out as much as I could."

His discussions with ORNL colleagues revealed they thought the chances were slim for cold fusion but that the laboratory should investigate it fully. Based on these weekend exchanges, the laboratory launched an intensified study of cold fusion beginning the following Monday. Teams in the Physics, Metals and Ceramics, Chemical Technology, and Engineering Physics and Mathematics divisions energized a dozen electrochemical cells to test the claims of cold fusion researchers, using more sensitive neutron-detection devices than those available to the purported discoverers of this energy source.

Michael Saltmarsh of the Fusion Energy Division chaired a laboratory committee compiling information on these experiments. Within a month, he testified before a U.S. House of Representatives science committee that the laboratory, despite its hot pursuit of cold fusion, had been unable to detect excess heat or radiation in its experiments. This and a bevy of similar reports from other laboratories discredited the University of Utah's claims. Frank Close, an ORNL scientist, subsequently published a critique of the short-lived cold fusion events, emphasizing the importance of following accepted scientific procedures when "new" phenomena are reported. Still, limited cold-fusion experimentation continued, in the United States, Europe, and Japan, in the hope that some yet-to-be explained phenomenon was occurring.

After a momentary cold shoulder, achieving magnetically confined hot plasma resurfaced as a major technological challenge at the laboratory and throughout the world of science. The pursuit, in fact, assumed cooperative global proportions during the 1980s, especially at the laboratory's large-coil testing stand named the International Fusion Superconducting Magnet Test Facility.

All major industrial nations conducted research on fusion power during the 1980s and on the superconducting magnets for use in fusion energy production. In cooperation with the International Atomic Energy Agency, DOE approved construction of a large magnetic coil facility at Oak Ridge to test huge superconducting magnets—three designed and fabricated in the United States by General Electric, General Dynamics, and Westinghouse and three overseas in Japan, Germany, and Switzerland. All used specifications written at ORNL so that the magnets would fit into the large-coil test facility.

The laboratory installed the six magnets, weighing forty-five tons each, in the toroidal (doughnut-shaped) large-coil facility. When its stainless steel vacuum chamber lid was lowered into place atop the magnets and the proper vacuum was achieved, its liquid helium refrigeration system chilled the magnets to almost absolute zero. Paul Haubenreich, assisted by Martin Lubell, managed comparative testing of the magnets during 1986 and 1987, checking their ability to withstand thermal, mechanical, and electrical stresses and determining whether superconducting coils were practical for confining the plasma of fusion reactors.

The large-coil stand operated reliably during twenty-two months of testing, and the magnets performed well, setting records as the largest superconducting magnetic coils in size, weight, and energy ever operated. It also marked the first time that four nations—the United States, Germany, Japan, and Switzerland—had submitted individual versions of similar equipment to collaborative testing for evaluation of their performance, reliability, and costs.

The 1988 report on the experiment stated that the magnetic coils in operation had exceeded their design parameters, indicating that much larger magnets could be built using similar design methods. The report observed that the successful international cooperation marking the large-coil tests boded well for other cooperative global ventures in fusion research.

These conclusions proved useful in the design of the International Thermonuclear Energy Reactor (ITER), planned as a joint effort of the United States, Russia, Japan, and the European community. This thermonuclear energy reactor was being conceived as the first fusion reactor in which studies of ignited and burning plasmas could be conducted.

Within the political and scientific communities of the United States, some observers recoiled at the costs of long-term fusion research, fearing that federal research funds would not be available for the long haul. After all, scientists projected that successful fusion energy generation would not occur until the mid-twenty-first century. "Let us not grow weary while doing good," warned William Happer, chief of DOE's Office of Energy Research. Quoting a letter from the Apostle Paul to the Galatians, Happer continued, "for in due season we shall reap if we do not lose heart."

The laboratory expected to play a significant role in the ITER program, and Paul Haubenreich, manager of the large-coil tests, went to

Europe for several years to work in that program. (Charles Baker of the laboratory now leads U.S. efforts in the ITER design.) After completing the large-coil tests, Martin Lubell and the laboratory's superconductivity team turned to potential commercial investigations of motors using superconducting materials. Their first superconducting device was in operation by 1990, and improvements in this device may result in the development of smaller, more efficient motors. ˙

Stellar Performance

Other advances in fusion energy research at ORNL during the late 1980s and early 1990s included improved plasma fueling and heating devices and construction and testing of an advanced toroidal facility, a stellarator fusion reactor shaped more like a cruller than the tokamak doughnut.

Pioneered by Stanley Milora and Christopher Foster at the laboratory, fueling fusion plasmas by freezing deuterium and later tritium into pellets and firing them into reactors became the standard fueling method worldwide. The laboratory became DOE's lead agency for this plasma fueling technology. For the ever-larger fusion reactors, the laboratory fabricated bigger pellets, discharging them into plasmas using an electron beam accelerator to vaporize their back ends and provide a rocketlike forward thrust. The laboratory also completed a radio frequency facility in 1985 to test the use of radio waves for heating of fusion plasmas, and it joined with Japan's energy institute to conduct collaborative testing at ORNL reactors of structural alloys that are candidates for fusion devices.

The laboratory also designed and built an advanced toroidal facility to supplant its impurities study experiment tokamak of the 1970s. Called a torsatron or stellarator, the advanced toroidal facility had a helical field for plasma confinement provided entirely by external coils, instead of relying on currents within the plasma as the tokamaks did. Aiming to create more stable plasmas, it afforded a steady, rather than a pulsed, operation, which utility systems prefer for electric power generation.

After four years of construction, the laboratory in 1988 completed its precision-crafted stellarator, with more than twice the plasma volume of previous stellarators. Its principal purpose was to determine the plasma

Under construction in 1986, the advanced toroidal facility for fusion research was completed in 1987.

pressure and stability limits for improved toroidal designs. Testing soon identified a second stability phase in the plasma, which was termed a major advance in fundamental plasma physics. The laboratory sought funding during 1992 for a restart and continued testing of this stellarator, which was the only fusion machine in the United States capable of operating in a steady state.

Although scientists had not achieved a self-sustaining controlled fusion reaction by 1993, the laboratory's fusion research inched ahead on research planks that scientists hoped would eventually build a steady platform for energy production. As Alvin Weinberg had warned decades before, however, for the moment and into the foreseeable future those planks would rest on an unsteady research base in a field that promised unlimited energy wrapped in the reality of endless problems. A three-minute burst of fusion energy at Princeton University in the fall of 1993 boosted the prospects for this intriguing energy source. But commercial applications of fusion energy remain decades away, at best.

Particle Accelerators

Global scientific cooperation is a two-way international highway. In the 1980s, the laboratory's Physics Division dispatched two of its large calorimeters and ten of its scientists to the European Laboratory for Particle Physics (CERN) in Switzerland to participate in experiments aimed at observing individual quarks outside nuclei.

The experiment fired oxygen nuclei into target nuclei of carbon, copper, silver, and gold at ultrahigh energies, dramatically demonstrating the conversion of energy into matter in a process that would have brought a smile to Albert Einstein's face. The laboratory calorimeter team, in effect, saw particles bombarding the gold nuclei multiply into many more particles.

Trivelpiece helped persuade the Reagan administration to explore these mysteries through funding for construction of a superconducting supercollider, a fifty-three-mile oval track to be built underground in Texas, where two opposing beams of protons would circle and collide. Seeking to determine whether quarks are the fundamental units of matter or if they can be further subdivided, this huge proton racetrack would have been the world's most powerful accelerator. Congress, however, voted to shut down the project in the fall of 1993, citing federal budget constraints and more pressing national concerns. Trivelpiece called the demise of the superconducting supercollider "a loss to science and the nation."

ORNL participation in the superconducting supercollider project involved developing detectors to determine the results of particle collisions. In 1989, the laboratory formed an Oak Ridge Detector Center, directed by Tony Gabriel. The center hoped to be at the forefront of developing central-system particle detectors for the supercollider that could track and measure the directions and initial energies of secondary particles produced by the collisions. Recognizing the value of these devices to global science, the laboratory consulted physicists from many nations for the detector designs. All of these efforts have been sidetracked by the demise of the superconducting supercollider.

Human Genome Initiative

Inspired by a U.S Congress, Office of Technology Assessment, report on detecting inherited mutations in human beings, DOE Office of Health and Environmental Research in 1987 launched an international campaign to map and sequence the 3 billion chemical base pairs in human DNA. Among the practical benefits of sequencing the human genome could be new diagnostic tests and therapies for genetic diseases such as cystic fibrosis, Huntington's disease, and Alzheimer's disease.

Through participation in long-term international studies of the survivors of the Hiroshima and Nagasaki bombs, ORNL researchers had obtained decades of experience in human gene studies that ultimately extended beyond their initial research into genetic abnormalities caused by intense radiation. In the 1970s, for example, the Biology Division had devised gene mapping techniques for the study of mutagens and carcinogens. Searching for genes that might inhibit cancer, they had identified individual genes and assigned them to specific chromosomes.

Laboratory capabilities were further enhanced by development during the 1980s of improved scanning tunneling microscopes that could obtain images of DNA strands. These microscopes could help determine the locations of genes on cell chromosomes (mapping) and the arrangement of DNA bases in the genes (sequencing) of the human genome. Sponsored by DOE and NIH, the human genome studies, an immense computer-intensive investigation, became global in scope, with several nations and many research facilities sharing the research and its costs.

The laboratory, however, had no externally funded human genome projects when the director of DOE health and environmental research programs visited Oak Ridge in 1990. Several internal seed money research projects set the stage for convincing DOE that the laboratory should be involved in the genome challenge. Six laboratory divisions subsequently participated in genome research, focusing on learning the order of chemical bases that make up DNA and locating specific genes to determine their functions.

Using mass spectrometry, gel electrophoresis, radiolabeling, laser ionization, and other research techniques, the laboratory sought information on the intricate human genome. It also provided a forum for international exchange of genome information in its Human Genome Management Information System, located in the Health Sciences Research Division.

The laboratory also offered a maze of knowledge about mouse genetics that might be useful to human genome researchers. DOE therefore encouraged collaboration between ORNL mouse experts and the genome centers. One outgrowth was a Biology Division initiative funded by the National Institute of Child Health and Human Development. The program, led by Richard Woychik and Gene Rynchik, used transgenic mice to help ascertain the locations and molecular structure of human genes. In time, this research could help advance understanding of human genetic disorders.

The laboratory, DOE, NIH, and, more generally, the international life sciences community hoped to obtain information on genes that would help them determine, for example, which genes are responsible for cystic fibrosis, Huntington's disease, and Alzheimer's disease. With this information, scientists eventually might devise methods for repairing these genetic disorders. Program advocates suggested that this information might also contribute to ameliorating mental health problems by identifying, for example, genetic causes of manic depression and schizophrenia. The most avid proponents asserted that successful completion of this global project could place humans in control of their genetic destiny, although critics questioned the wisdom and ethics of this goal.

Environment, Safety, and Health

In June 1989, Admiral Watkins outlined a "new culture of accountability" for DOE to regain its credibility in environmental restoration compliance and, more generally, the overall management of its facilities. To help reach this goal, in an unprecedented move the admiral permitted state agencies access to DOE installations to monitor DOE compliance with environmental standards and regulations. DOE also emphasized environmental, safety, and health compliance in awarding fees to contractor-operators of its facilities, mandated full compliance with federal Occupational Safety and Health Administration (OSHA) standards,

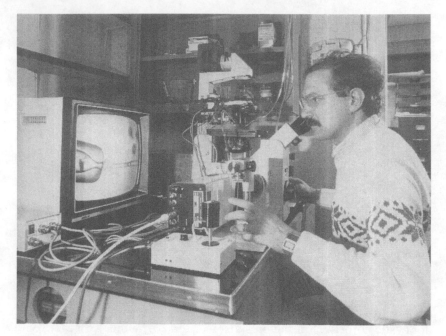

Richard Woychik injects foreign DNA into mouse eggs as a tag for locating the structure and function of the genomic region.

and formed "tiger teams" to assess field agency compliance and corrective measures.

These measures were a belated response to the 1984 amendments to the Resource Conservation and Recovery Act (RCRA) stipulating that facilities (including those sponsored by the federal government) that handled hazardous wastes must reduce the generation of such wastes and remediate sites where these wastes already existed. It soon became apparent that remediating hazardous wastes would be a time-consuming and costly enterprise, and that there would be no cheap, quick fix. John Gibbons, director of the Office of Technology Assessment, who previously had worked at both the laboratory and the University of Tennessee's Energy, Environment, and Resources Center, declared, "Decades will be required for cleanup of certain sites while others will never be returned to pristine conditions."

As an incentive to reduce wastes, the laboratory adopted a charge-back policy, billing waste-disposal costs to the division that generated

The laboratory improved this remotely operated vehicle for the mapping and measurement of contaminated waste sites.

the waste. Laboratory research and development proposals would now have to incorporate waste disposal into their estimated project costs, encouraging researchers to avoid using toxic substances in their experiments. "It's a new mentality, a cultural change," Thomas Row insisted. Row, who in 1991 became director of the Office of Environmental, Safety, and Health Compliance, concluded that major changes in ORNL waste-disposal methods were a reflection of the laboratory's new corporate culture born of a new age of environmental accountability.

Historically, for example, the laboratory had placed solid low-level hazardous and radioactive wastes, such as contaminated glass and cloth, into unlined trenches; now it packaged such waste in steel cans placed inside concrete vaults that were eventually entombed in earth berms equipped with monitored drainage systems. Low-level liquid wastes, once disposed of using underground hydrofracture, were now concentrated and compacted to reduce the volume, then solidified and stored aboveground. The laboratory's high-level spent reactor fuel went to the

Idaho or Savannah River complexes, which had storage facilities for re-actor fuel that required reprocessing. The laboratory's transuranic wastes were stored on-site in specially designed bunkers for eventual disposal at a DOE centralized facility, perhaps the Waste Isolation Pilot Plant in New Mexico. One measure of the laboratory's commitment to environ-mental, safety, and health programs was its increase of program per-sonnel from 240 in 1988 to 390 in 1990.

In 1988, about 15 percent of the ORNL budget was devoted to waste-management and remedial actions—and this was only the beginning. To reduce waste-management and remediation program costs, the national laboratories were challenged to find ways to treat the contamination without moving it. One ORNL response involved in situ vitrification, which entailed passing electric currents underground through hazard-ous wastes, heating them to high temperatures, and thereby convert-ing them into glasslike solids impervious to groundwater.

Developed at the Pacific Northwest Laboratory, *in situ* vitrification was tested by Brian Spalding and colleagues at ORNL to isolate stron-tium and cesium. Although still an expensive technique, *in situ* vitrifi-cation may be used at some future date to treat the pits and trenches that served as waste repositories during the laboratory's early years.

Another innovation was bioremediation, which uses microorganisms to break down hazardous chemicals. ORNL teams used methane-consum-ing microorganisms in the soil to break down gasoline and other solvents. Additional research was under way in 1993 to identify or modify micro-organisms to consume other types of toxic wastes.

All of these measures and strategies were part and parcel of a broad research effort to help resolve environmental problems that had been neglected for too long. Waste management, in short, now promises to be a major focus of the laboratory's agenda during the next half-century.

Tigers on the Prowl

To ensure full compliance with environmental, health, and safety pro-grams, Admiral Watkins dispatched "tiger teams" to DOE field orga-nizations for thorough operational and management inspections. With a tiger team ready to pounce on Oak Ridge in 1991, the laboratory instituted a massive cleanup in advance of the inspection. Its Plant and Equipment Division, for example, installed more than 1,500 new safety guards on 574 machines. Within a month after the lengthy inspection, the laboratory's response team had corrected 366 deficiencies identified by the tigers.

Complimenting William Fulkerson, Jerry Swanks, Frank Kornegay, William Morgan, Tony Wright, and David Reichle on their leadership of the laboratory's lionhearted response, Trivelpiece declared the tiger team inspection largely a success, although he also pointed out that the laboratory had not received a completely clean bill of health.

"We did not come through unscathed," he admitted. "There are a lot of problems: legacies from past practices, deficiencies in current programs, and management deficiencies and improving acceptance for environmental safety and health." Nevertheless, he thought the tiger team inspection had served as a catalyst for progress.

Defense Challenges

Visiting the laboratory in 1992, President Bush referred to the facility as an "arsenal of democracy." Although a scientific rather than a weapons laboratory, ORNL had supported national defense at every opportunity. In addition to assisting the Strategic Defense Initiative, the laboratory undertook research during the 1980s for the Defense Department that included investigations of defense materials, battlefield logistics, robotics, instruments and controls, and electromagnetic interference.

Also for the Defense Department, the laboratory's radioisotopes group directed by Neil Case developed isotope-powered lights using radioactive emissions from krypton and tritium to excite phosphor pellets, causing them to glow in the dark. These "plugless" lights provided landing and distance markers for military and civilian pilots in remote areas. In another case, Cabell Finch and Lynn Boatner of the Solid State Division developed doped crystals for room-temperature promethium lasers. These crystals were suited for satellite-to-submarine underwater communications because their light can be transmitted through space and water.

Led by the Energy Division's Samuel Carnes, in 1987, an ORNL team completed the final programmatic environmental impact statement for disposal of the army's stockpiled chemical weapons. The team identified on-site incineration as the environmentally preferred method of weapons-disposal management.

When terrorists bombed aircraft during the 1980s, a team in the Analytical Chemistry Division devised an explosives sniffer using mass spectrometry to test the air for suspect chemicals, thereby determining in seconds whether explosives were present. This interested airport security firms, and Energy Systems licensed the "sniffer" to a private company for commercial use.

In addition, the laboratory developed a direct-sampling ion-trap mass spectrometer. Installed in a van, this equipment served as the basis of a mobile laboratory for rapid detection of organic contaminants at environmental cleanup sites. Providing test results much faster than conventional methods, the device is expected to produce substantial savings for the waste cleanup programs of both DOE and the Army Toxic and Hazardous Materials Agency.

Computer models developed by the laboratory's Center for Transportation Analysis in the Energy Division saw useful application during the 1991 Gulf War. The U.S. Transportation Command used the software to schedule deployment of troops and equipment to the Middle East for Operations Desert Shield and Desert Storm in the largest airlift operation in history.

Despite the wide range of defense-related initiatives pursued at the laboratory, successful "national defense" in the post–Cold War world ultimately will rest on economic prosperity. To meet this challenge, during the 1990s the laboratory increasingly focused its resources and staff on environmental and economic, not military, matters. The key words for this operation were technology transfer and national competitiveness.

Technology Transfer

ORNL efforts to transfer its technological advances to industry began in 1962, when Alvin Weinberg established an Office of Industrial Cooperation to reduce the time required for the civilian economy to adopt scientific advances. Carol Oen and Don Jared headed the laboratory's "technology utilization" offices during the 1970s and found partial suc-

Mike Guerin and Mark Wise work at an ion-trap mass spectrometer used at the laboratory to identify pollutants in the environment.

cess through spin-off industries often launched by former ORNL personnel. The laboratory and other Energy Systems sites also helped lure Science Applications, System Development, TRW, Exxon, Bechtel, and other corporations to Oak Ridge by increasing outside awareness of local technical capabilities and opportunities for profit.

Legal barriers involving patents and nonexclusive licensing, however, hampered quick technological transfer. Corporate executives were reluctant to invest in technology without the marketplace advantage of holding the exclusive rights to a particular technology.

Recognizing these difficulties and the frustration of industry, DOE initiated a technology transfer pilot project centered around newly discovered high-temperature superconductors with the intention of streamlining legal requirements at DOE laboratories. Oak Ridge, Los Alamos, and Argonne national laboratories were designated as Superconductivity Pilot Centers, and the three worked closely with DOE, under industry's watchful eye, to devise procedures that would accelerate the transfer of superconducting technology from these laboratories to industry. ORNL's Tony Schaffhauser, Louise Dunlap, and Jon Sodestrom, in cooperation with their counterparts in other national laboratories, thus helped establish the procedures that became a model for DOE CRADAs legislated by Congress.

Recognizing the latent economic power in the national laboratories, Congress passed legislation during the 1980s to encourage technology transfer. The laboratory's new contractor operator, Martin Marietta Energy Systems, vigorously promoted this initiative. In 1985, for example, Energy Systems signed an exclusive license with Cummins Engine Company for the use in diesel engines of modified nickel aluminide alloys developed by ORNL's C. T. Liu and his colleagues. Energy Systems offered financial incentives to ORNL personnel who applied for patents as well. ORNL inventors received the first royalties for their innovations in 1987.

Direct cooperation with industry at the Roof Research Center, High Temperature Materials Laboratory, and elsewhere quickened the pace of transferring information on ceramics, semiconductors, electronics, computer software, insulation, and other commercially promising technologies. As a result, the laboratory led other DOE facilities in technology transfer, and its program became a model for other agencies to emulate.

Industry, for example, expressed great interest in the laboratory's development of ceramic gel casting and material reinforcement with whiskers made from silicon carbide. By 1989, eleven companies had obtained licenses to use durable whisker-toughened ceramic composites in metal-cutting tools. A license for using gel casting to shape ceramics, a process invented in the Metals and Ceramics Division, went to Coors Ceramics Corporation. Coors built a plant in Oak Ridge to pursue this and related technologies.

Trane Company, a worldwide manufacturer of air-conditioning and refrigeration systems, acquired a license for gas-powered absorption chillers invented at the laboratory by Robert DeVault. These gas chillers were more economical and efficient than electric chillers and could reduce summer demands for electricity by shifting commercial air-conditioning loads to natural gas.

Energy Systems issued its first royalty-bearing license in nuclear medicine to Du Pont in 1989. Prem Srivastava and associates in the Health and Safety Research Division synthesized a chemical compound useful for cancer detection that Du Pont expected to market to medical research institutions.

Open for Business

The 1989 Technology Transfer Act, passed by Congress, amended the Atomic Energy Act to make technology transfer a principal mission of DOE and its laboratories. The act allowed contractor-operated laboratories such as Oak Ridge to work directly with industries, universities, and state governments, jointly sponsoring research and sharing information through CRADAs. "The labs are now open for business," proclaimed William Carpenter, Energy Systems' chief of technology transfer.

In 1990, the laboratory entered its first CRADA, joining an international consortium to study chemicals that could serve as alternatives to CFCs. More CRADAs followed. During his February 1992 visit to the laboratory, President Bush highlighted this technology-transfer program at the signing of a CRADA with Coors Ceramics to develop precision machining of ceramics.

Touring the High Temperature Materials Laboratory and addressing a crowd in front of the building, President Bush praised the $3.6-million CRADA with Coors as an excellent way to take technology directly to markets and create new jobs. "The High Temperature Materials Laboratory is a world-class facility," he declared, "and in the race with other nations in making precision parts, America will get there first." After Trivelpiece and Joseph Coors signed the CRADA, Coors presented the president with a ceramic golf putter as a lighthearted example of the products that could flow from the materials research.

Speaking in Knoxville later that day, the president promised significant increases in federal funding for science education and NSF. Thus, the president's brief visit to Oak Ridge and Knoxville framed, in national terms, two of the laboratory's most important initiatives of the 1990s—technology transfer and science education.

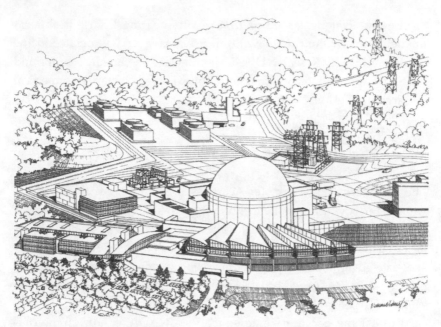

Architectural rendering of the proposed Advanced Neutron Source (ANS).

Future Challenges

In listing the future priorities of the "broadest based and most multidisciplinary of the DOE national laboratories," Alvin Trivelpiece highlighted his hope that the laboratory will become the center for excellence in research reactors. Its high-flux isotope and tower shielding reactors were now back in service, although the latter was funded primarily by a Japanese-sponsored program for breeder reactor shielding studies and was shut down in late 1992.

In 1992, the laboratory continued to press for funding to design and construct a major new reactor to replace its aging high-flux isotope reactor. Named the advanced neutron source (ANS), this proposed reactor was first studied in 1984 as a Director's Fund project; the first funding from DOE arrived in 1987.

Leadership in neutron-scattering research had passed from the United States to Europe during the 1970s when a reactor was built at Grenoble, France, that had a neutron flux and experimental facilities superior to those at Oak Ridge and other DOE laboratories. Backed by reports of several important national committees, ORNL management contended that building the ANS would regain world leadership for the United States by providing the most intense steady-state neutron beams in the world.

Managed initially by Ralph Moon and David Bartine, the project was later placed under Colin West, who directed plans for the conceptual design of the ANS with the aid of prominent scientists throughout the world. Surrounding the reactor would be a national research center, with adjoining structures housing minilaboratories for neutron-scattering and other experiments, as well as offices for scientists from both the laboratory and elsewhere.

The initial reactor design called for heavy water to cool the fuel and reflect neutrons back into the core. (The high-flux isotope reactor used ordinary water as coolant and beryllium as a neutron reflector.) Studies by the laboratory and the Idaho National Engineering Laboratory led to the selection of a split-core configuration with uranium silicide fuel in aluminum-clad plates. This would permit a 200- to 350-megawatt powerhouse, compared with the original 100-megawatt rating of the high-flux isotope reactor.

Noting that the laboratory had built and operated fourteen nuclear reactors (counting the 1955 Geneva conference reactor and the pool critical assembly), Murray Rosenthal observed that this ANS would become ORNL's fifteenth reactor and the first one built since 1966. Colin West estimated its 350 megawatts and modern beam facilities would provide neutron beams with intensities at least ten times those of the high-flux isotope reactor and at least ten thousand times greater than those available to Wollan and Shull at the 1943 graphite reactor. John Hayter, scientific director for the project, said the new reactor's plans would include about thirty beam lines and beam guides, many of which would serve more than one instrument. There would be special features such as neutron mirrors for beam delivery and two cold sources (tanks of liquid deuterium) to slow some of the neutrons before they are transported to the "guide hall" for experiments.

The conceptual design involved personnel from national laboratories, industries, and universities, plus researchers from Germany, Japan, and Australia. More than a thousand non-ORNL scientists are expected to conduct research annually at the ANS when it becomes operational. With the aging high-flux isotope reactor operating at reduced power to prolong its life, early completion of the ANS seemed vital. "When the HFIR reaches the end of its useful life, we will need a new reactor to enable U.S. scientists to conduct neutron-scattering studies to make progress in certain key fields," Trivelpiece asserted. "I think," he added, "we need to make a full court press, and I regard this project as the highest-priority technical facility pursued by the Laboratory."

Other ongoing reactor programs at the laboratory include the modular high-temperature gas-cooled reactor research program, which is seeking to develop a smaller-scale and inherently safer reactor than those currently in use. The laboratory is also providing research and design review support for the liquid-metal fast breeder reactor with potential for greatly extending the nuclear fuel supply, and it is reviewing and researching DOE's work with improved light water reactors of modular size and improved safety characteristics. ORNL support for safety studies and research for NRC is expected to continue as well.

Established in 1943 as a nuclear reactor, chemical separations, and scientific laboratory, ORNL continued to build upon these traditional strengths in 1993. Nevertheless, broadening investigations into other energy forms, into issues of environmental safety and health, and into technical advances that sought to improve national economic competitiveness absorbed ever-larger portions of the laboratory's budget and resources as it approached the end of the twentieth century.

The laboratory's future, in fact, seemed to lie not so much in its ability to do research in specific nuclear projects as in its deeply rooted skills to undertake large-scale, complicated projects that addressed broad national and international concerns. How well it performs these services in a variety of energy and environmental fields could well determine the laboratory's future. These efforts, in turn, could help chart America's future, enabling the nation to retain center stage in an increasingly complicated and competitive world.

Epilogue

Nearing retirement in 1992 after more than fifty years of service to ORNL, senior staff advisor Donald Trauger reflected on the lessons of a half-century. The laboratory and science at large, he urged, should expand their strategic planning to longer time spans. "Recent changes of national administration have limited effective implementation of some programs to four years or even two years, and industry is shortening its planning to as little as two years because of high capital costs and demands for early returns on investments. Perhaps," Trauger suggested, "the national laboratories can effectively consider the time spans that are really desirable. Even 100 years is not as distant as we might have thought."

ORNL management, as always, devotes considerable attention to planning the institution's future research and acquiring the equipment and facilities needed to support world-class science. DOE, in fact, requires the laboratory to prepare institutional plans looking five years into the future, and in 1990 laboratory director Alvin Trivelpiece formed a planning group to analyze the laboratory's long-term corporate strategy.

In addition to assigning the highest priority to the laboratory's proposed advanced neutron source (ANS) and other nuclear reactor studies, Trivelpiece emphasized the global importance of the work of Milton Russell and the Energy Division teams, who by 1992 had assisted twenty-one nations with the development of their energy and environmental technology policies. Pointing out that events in these nations had worldwide environmental ramifications, Trivelpiece urged Congress to support ORNL efforts to assist other nations in meeting their energy needs while reducing the strains on the environment and world oil markets.

In recent years, the laboratory hosted elementary school children interested in science.

To improve science education, Trivelpiece advocated greater cooperation with Oak Ridge Associated Universities, Pellissippi State Community College, Roane State Community College, the University of
Tennessee, Tennessee state government, and regional and local school
systems. He was particularly interested in designing classrooms for the
twenty-first century using reasonably priced electronic teaching aids
and student workstations.

For the 1990s and beyond, Trivelpiece and the strategic planning
group expected the laboratory's directions to be dominated by four
major themes: education, energy, environment, and economic competitiveness. They proposed to support these efforts with three new major
user facilities that the laboratory hopes to complete within a decade:
the ANS and its adjoining research facilities, a materials science center
on the laboratory's east end, and an environmental and life sciences
center on the west end.

Demonstrating the increasing importance of materials science to
the laboratory's efforts to improve national economic competitiveness
is the proposed Materials Science and Engineering Complex, which the

View of the modern ORNL with the Holifield Heavy Ion Research tower in the fore-
ground.

laboratory hopes to construct near the Holifield Heavy Ion Research
tower. Consolidating existing programs in new facilities to enhance scientific interaction, the complex would include centers for solid-state research and processing, advanced materials research, and composite
materials investigations.

Explosive growth in the materials sciences—along with their role in
ORNL's technology-transfer programs since 1980—has severely over-
crowded existing laboratories. Building a new complex is considered
more economical than upgrading older structures to meet today's environmental and safety standards. The proposed new complex aims to enhance the on-site involvement of university and industrial researchers in
cooperative ceramics, composites, superconductors, and high-temperature metals and alloys projects. This complex, not surprisingly, enjoys
strong support from universities and industries in the Southeast.

At the western gate, near the existing Environmental Sciences and
Aquatic Ecology laboratories, the laboratory proposes to develop an
Environmental, Life, and Social Sciences Complex. The complex would

include centers for biological sciences and earth systems and a biological imaging and advanced photonics laboratory. Its completion would place under one roof ORNL research in structural biology, biotechnology, genetics, global environmental studies, risk assessment and management, environmental restoration, the social sciences, energy efficiency, and transportation systems.

As with the materials science divisions, research in the environmental, life, and social sciences in 1992 was scattered throughout the laboratory in older facilities. For example, the Biology Division had been housed since 1946 in obsolete facilities at the Y-12 plant, eight miles from the X-10 complex. With much of the laboratory's global research centered in the newly formed Environmental, Life, and Social Sciences Directorate, the collaborative interactions facilitated by concentrated research in this new complex would help open new horizons for the solution of global challenges.

One noteworthy problem area for the laboratory lies in nuclear physics. Although the Holifield heavy-ion research accelerator was only twelve years old in 1992, and had set new records for beam energies in 1992 that were 60 percent higher than those achieved in 1982, it had fallen on hard times. New European accelerators provided even higher energies, and budget cuts reduced the operating time available for researchers at the Holifield accelerator.

To reverse these trends, the laboratory proposed to use the Holifield facility to accelerate radioactive ion beams, a unique capability that would extend the facility's value to nuclear physicists, especially those interested in astrophysics. If this proposal is approved, a recoil mass spectrometer, jointly funded by the laboratory and universities, would be acquired to complement the radioactive beam capability.

While applying cost constraints to facilities such as the Holifield accelerator, DOE began to devote vast resources during the 1990s to improving scientific understanding of the transport of wastes in the environment and the remediation of waste-disposal sites. As a result, Alvin Trivelpiece expected the laboratory to greatly expand its waste-management and remediation work. Such efforts would not only help stabilize—and perhaps reverse—the environmental damage caused by previous projects but also would help regain public confidence in the institution. Revelations about past environmental abuses, combined with the laboratory's initial reluctance to acknowledge the problem, have made the public increasingly wary about both the laboratory's work and its word. Successfully addressing the waste problem could persuade the public that the laboratory is cleaning up its act and is again worthy of citizen confidence. Thus for all the concern about the environment and public health, the greatest beneficiary of an effective cleanup program could be the laboratory itself.

Back to the Future

As the laboratory approached its second half-century, Alvin Weinberg was busy in retirement preparing Wigner's papers for publication. His effort to uncover and organize the laboratory's past gave Weinberg an opportunity to reflect on Wigner's legacy.

Weinberg observed that Wigner, the laboratory's most renowned scientist, not only set a standard of performance for Oak Ridge but offered a vision of the future that speaks as directly to the apprehensions of the 1990s as it did to the uncertainties of the 1940s. In a simple statement of truth, Wigner once remarked, "Every moment brings surprises and unforeseeable events—truly the future is uncertain."

Weinberg himself viewed aging and the future with equanimity. He wryly concluded that scientists improve with age because their knowledge broadens as they become older. Much of science, he said, comes not out of brilliant flights of fancy but from viewpoints and techniques growing out of a lifetime of scientific inquiry. The same sentiment might well apply to an institution that reaches the half-century mark. Its corporate experience and accomplishments should serve as a foundation of strength upon which to build a vigorous future of inventiveness and purpose.

Although the laboratory, like science generally, seems more interested in the future than the past, it sometimes turns to its past for hope, inspiration, and understanding. When drafting plans for ORNL initiatives during the 1990s, the strategic planning group admitted that improving national competitiveness through technology transfer and science education might be "more difficult than the Manhattan Project, which birthed the national laboratories nearly a half century ago."

In the 1940s, the nation's attention and resources were riveted on winning the war, and ORNL efforts on behalf of the atomic bomb received the highest priority. Today, the enemies are less clearly defined and ORNL initiatives must share the political and budget spotlight with other government priorities and needs. Thus, the laboratory will have to work even harder to justify public investment in its research activities. As Weinberg recently suggested, if the laboratory is to become a prime engine of the national economy, its people must "adopt the same high standards and dedication shown during the four years of the Manhattan Project."

And so the experience of the laboratory has come full circle. Amid the complex of buildings, intricate equipment, roads, reactors, accelerators, robots, piping, and computers, one force stands above all others in explaining the institution's success: the dedication and commitment of the people who work there. Ironically, that dedication and commitment reached its first peak during the war years, when secrecy prevailed. Fifty years later, the laboratory is determined to open its doors to the future, drawing on the storehouse of knowledge and skills found among its staff, to serve the public interest. Whether the laboratory can generate the same level of commitment from its workers that characterized the years of the Manhattan Project remains to be seen. Recent lawsuits by Charles Varnadore, who accused laboratory management of harassment for his bringing safety concerns to their attention (accusations upheld by a Department of Labor administrative law judge), did little to help laboratory efforts to convince others that it was creating a new institutional culture of openness.

Despite such internal workplace controversies, this much is certain: The purposes to which the laboratory can now apply its talents are more diffuse and difficult to pinpoint. But for ORNL, the future has always been uncertain and unclear. Its staff has seized opportunities and redefined the laboratory's purposes time and again to fit changing circumstances.

As the laboratory celebrates its fiftieth anniversary and as it stands on the threshold of the twenty-first century, there is little doubt that it will marshal its resources and talents to meet the challenges of tomorrow. At the dawn of a new era, ORNL history foretells this much about the future: If the laboratory's past is its prologue, then its next fifty years should be as demanding, rewarding, and surprising as its first half-century. That simple truth is likely to make its future as worthwhile to watch as its history has been to study.

Bibliography

Interviews

Abee, Harold H. Interviewed by Susan M. Schexnayder. Oak Ridge. 29 Oct. 1991.

Adamson, George. Interviewed by Susan M. Schexnayder. Oak Ridge. 6 Nov. 1991.

Auerbach, Stanley. Interviewed by Susan M. Schexnayder. Oak Ridge. 11 Oct. 1991.

Borkowski, Casimer J. Interviewed by Susan M. Schexnayder. Oak Ridge. 16 Oct. 1991.

Cardwell, David. Interviewed by Susan M. Schexnayder. Oak Ridge. 30 Oct. 1991.

Chester, Conrad. Interviewed by Susan M. Schexnayder. Oak Ridge. 3 Sept. 1991.

Cohn, Waldo. Interviewed by Leland R. Johnson. Oak Ridge. 11 July 1991.

Cole, Thomas. Interviewed by Susan M. Schexnayder. Oak Ridge. 15 Jan. 1992.

Culler, Floyd L. Interviewed by Daniel Schaffer. Palo Alto, Calif. 23 July 1991.

Cunningham, John E. Interviewed by Susan M. Schexnayder. Oak Ridge. 24 July 1992.

Dabbs, John W. Interviewed by Susan M. Schexnayder. Oak Ridge. 19 May 1992.

Fermi, Rachael. Interviewed by Leland R. Johnson. Oak Ridge. 6 Aug. 1991.

Foster, Ward E. Interviewed by Susan M. Schexnayder. Oak Ridge. 13 July 1992.

Fox, Richard. Interviewed by Leland R. Johnson. Oak Ridge. 6 Aug. 1991.

Gailor, JoAnn. Interviewed by Susan M. Schexnayder. Knoxville, Tenn. 26 Nov. 1991.

Gillette, John H. Interviewed by Leland R. Johnson. Oak Ridge. 20 June 1991.

Hurst, Fred. Interviewed by Susan M. Schexnayder. Oak Ridge. 7 July 1992.

Johnson, Linton C. Interviewed by Susan M. Schexnayder. Oak Ridge. 17 June 1992.

Keim, Christopher P. Interviewed by Susan M. Schexnayder. Oak Ridge. 14 Jan. 1992.

Leverett, Miles. Interviewed by Leland R. Johnson. Oak Ridge. 23 July 1991.

Levy, Henri. Interviewed by Susan M. Schexnayder. Oak Ridge. 9 July 1992.

Lietze, Milton. Interviewed by Leland R. Johnson. Oak Ridge. 8 Aug. 1991.

Lincoln, Thomas. Interviewed by Susan M. Schexnayder. Farragut, Tenn. 15 Oct. 1991.

Livingston, Robert. Interviewed by Susan M. Schexnayder. Knoxville, Tenn. 5 Mar. 1992.

MacPherson, H. G. Interviewed by Susan M. Schexnayder. Oak Ridge. 22 July 1991.

McCullough, Frederick C. Interviewed by Susan M. Schexnayder. Oak Ridge. 3 Mar. 1992.

Pomerance, Herbert. Interviewed by Susan M. Schexnayder. Oak Ridge. 4 Sept. 1991.

Postma, Herman. Interviewed by Susan M. Schexnayder. Oak Ridge. 24 Sept. 1991.

Ramsey, Mansel. Interviewed by Susan M. Schexnayder. Knoxville, Tenn. 17 Mar. 1992.

Richmond, Chester A. Interviewed by Susan M. Schexnayder. Oak Ridge. 21 May 1992.

Robinson, Mark. Interviewed by Leland R. Johnson. Oak Ridge. 23 July 1991.

Stoughton, Ray W. Interviewed by Leland R. Johnson. Oak Ridge. 8 Aug. 1991.

Taylor, Ellison. Interviewed by Leland R. Johnson. Oak Ridge. 11 July 1991.

Terry, Michael. Interviewed by Susan M. Schexnayder. Oak Ridge. 24 Aug. 1992.

Trauger, Donald B. Interviewed by Leland R. Johnson. Oak Ridge. 6 Aug. 1991.

Weinberg, Alvin. Interviewed by Leland R. Johnson. Oak Ridge. 25 Sept. 1991.

Wilkinson, Michael. Interviewed by Leland R. Johnson. Oak Ridge. 18 July 1991.

Winters, Charlie. Interviewed by Donald B. Trauger. Bethesda, Md. 8 July 1991.

Articles

Aftergood, Steven, David Hafemeister, Oleg Prilutsky, Joel Primack, and Stanislav Rodionov. "Nuclear Power in Space." *Scientific American* 264 (June 1991): 42–47.

Alexander, Tom. "The Hot Promise of Thermonuclear Power." *Fortune*, June 1970, 92–97, 126–34.

Anderson, Herbert. "The Legacy of Fermi and Szilard." *Bulletin of the Atomic Scientists* 30 (Sept. 1974): 56–62.

Anderson, Norman. "Zonal Centrifuges and Other Separation Systems." *Science* 154 (7 Oct. 1966): 103–12.

Anderson, Norman, and James Liverman. "Our First Priority in Science Should Be the Molecular Anatomy of the Cell." *Scientific Research*, 27 May 1968, 37–46.

Andrews, Peter. "In Chemistry, It's Where the Action Is." *Mosaic* 20 (Summer 1989): 36–43.

Allison, Samuel. "Arthur Holly Compton, Research Physicist." *Science* 138 (16 Nov. 1962): 794–97.

Atomic Energy Commission. Isotopes Branch. "The Isotope Distribution Program." *Science* 106 (29 Aug. 1947): 175–79.

Auerbach, Stanley. "The Soil Ecosystem and Radioactive Waste to the Ground." *Ecology* 39 (1958): 527.

———. "Environmental Issues in Domestic Energy Development." *Journal of the Tennessee Academy of Science* 53 (Apr. 1978): 43–44.

Auerbach, Stanley, John Trabalka, and Dean Eyman. "Analysis of the 1957–1958 Soviet Nuclear Accident." *Science* 209 (18 July 1980): 345–51.

Bacher, Robert. "The Development of Nuclear Reactors." *Bulletin of the Atomic Scientists* 5 (Mar. 1949): 80–94.

Ball, James, and John Pinajian. "The Electronuclear Division: Past and Future." *Oak Ridge National Laboratory Review* 5 (Fall 1971): 1–11.

Barkenbus, Jack, Alvin Weinberg, and Michael Alonso. "Storing the World's Spent Nuclear Fuel." *Bulletin of the Atomic Scientists* 41 (Nov. 1985): 34–37.

Bernstein, Barton. "A Postwar Myth: 500,000 U.S. Lives Saved." *Bulletin of the Atomic Scientists* 42 (June 1986): 38–40.

Berry, Stephen. "The Federal Laboratories." *Bulletin of the Atomic Scientists* 40 (Mar. 1984): 21–25.

Blake, Michael. "Fusion '84: Yet Another Redirection." *Nuclear News* 27 (June 1984): 133–38.

———. "Fusion '85: And the JT-60 Makes Three." *Nuclear News* 28 (June 1985): 77–83.

———. "Charting a Course for the National Laboratories." *Nuclear News* 34 (June 1991): 59–60.

Bonner, W. P., Tsuneo Tamura, C. W. Francis, and J. W. Amburgey, Jr. "Zonal Centrifugation—A Tool for Environmental Studies." *Environmental Science and Technology* 4 (Oct. 1970): 821–24.

Bronfman, Lois. "Setting the Social Impact Agenda: An Organizational Perspective." *Environmental Impact Assessment Review* 11 (1991): 69–79.

Brown, George, Jr. "Courting Disaster in Orbit." *Bulletin of the Atomic Scientists* 45 (Apr. 1989): 7–9.

Burwell, Calvin, M. J. Ohanian, and Alvin Weinberg. "A Siting Policy for an Acceptable Nuclear Future." *Science* 204 (June 1979): 1043–51.

Cockcroft, John. "Peaceful Uses of Atomic Energy: The Second International Conference." *Bulletin of the Atomic Scientists* 15 (Jan. 1959): 18–21.

Cohn, Waldo. "Mankind, the Beneficiary." *Monsanto Magazine* 25 (Dec. 1946): 35–37.

Cohn, Waldo, George Parker, and Edward Tompkins. "Ion-Exchangers to Separate, Concentrate and Purify Small Amounts of Ions." *Nucleonics* 3 (Nov. 1948): 22–33.

Compton, Arthur. "The Birth of Atomic Power." *Bulletin of the Atomic Scientists* 9 (Jan. 1953): 10–12.

Coutant, Charles. "Thermal Niches of Striped Bass." *Scientific American* 255 (Aug. 1986): 98–104.

Crease, Robert. "The History of Brookhaven National Laboratory: Part One." *Long Island Historical Journal* 3 (Summer 1991): 167–87.

———. "The History of Brookhaven National Laboratory: Part Two." *Long Island Historical Journal* 4 (Spring 1992): 138–61.

Cunningham, Burris. "Ultramicrochemistry." *Scientific American* 190 (Feb. 1954): 76–82.

Daniels, Farrington. "Plans and Problems in Nuclear Research." *Science* 104 (2 Aug. 1946): 91–96.

Dean, Gordon. "From Chain Reaction to Breeding." *Bulletin of the Atomic Scientists* 9 (July 1953): 218, 226.

de Laguna, Wallace. "What Is Safe Waste Disposal?" *Bulletin of the Atomic Scientists* 15 (Jan. 1959): 35–43.

de Solla Price, Derek. "The Beginning and End of the Scientific Revolution: 1670–1970." *Lehigh Alumni Bulletin*, Mar. 1961, 6–9.

EBT Group. "ELMO Bumpy Torus Programme." *Nuclear Fusion* 25 (Sept. 1985): 1271–74.

Edlund, M. C., and P. F. Schutt, "The Future of Thermal Breeders." *Nucleonics* 21 (June 1963): 76–78.

"Exploring the New Material World." *Science* 252 (3 May 1991): 644–45.

"Extraordinary Atomic Plane: The Fight for an Ultimate Weapon." *Newsweek*, 4 June 1956, 55–60.

"Former Secretary Turns Her Hand to Research." *Martin Marietta Today*, no. 1 (1991): 16.

"From the High School on the Mall to the Nobel Prize." *Elet es Irodalom* 7 (21 Dec. 1963): 7.

Fulkerson, William, David Reister, Alfred Perry, Alan Crane, Don Kash, and Stanley Auerbach. "Global Warming: An Energy Technology R&D Challenge." *Science* 246 (17 Nov. 1989): 868–69.

Gibbons, John H., Peter Blair, and Holly Gwin. "Strategies for Energy Use." *Scientific American* 261 (Sept. 1989): 136–43.

Gibbons, John H., R. L. Macklin, P. D. Miller, and J. H. Neiler. "Average Radiative Capture Cross Sections for 7- to 170-kev Neutrons." *Physical Review* 122 (1 Apr. 1961): 182–200.

Glass, Bentley. "Information Crisis in Biology." *Bulletin of the Atomic Scientists* 18 (Oct. 1962): 6–12.

"Goodbye Holifield, Hello Oak Ridge." *Science* 191 (6 Feb. 1976): 449.

Green, Harold P. "The Peculiar Politics of Nuclear Power." *Bulletin of the Atomic Scientists* 38 (Dec. 1982): 59–65.

"Greenhouse Economics: Count before You Leap." *Chemtech* 21 (Oct. 1991): 584–87.

"Greetings from Holifield National Laboratory." *Science* 187 (24 Jan. 1975): 240.

Hafstad, Lawrence. "Atomic Power for Aircraft." *Bulletin of the Atomic Scientists* 5 (Nov. 1949): 309–12.

———. "The Breeder and the Homogeneous Reactor." *Bulletin of the Atomic Scientists* 6 (Feb. 1950): 37, 49.

Hamilton, Joseph H., and J. A. Maruhn, "Exotic Atomic Nuclei." *Scientific American* 255 (July 1986): 80–89.

Hammond, R. Philip. "Large Reactors May Distill Sea Water Economically." *Nucleonics* 20 (Dec. 1962): 45–49.

———. "Low Cost Energy: A New Dimension." *Science Journal*, Jan. 1969, 34–44.

Hayter, John, John Axe, and Roger Pynn. "Neutrons and Materials Science." *MRS Bulletin* 15 (Nov. 1990): 42–47.

Henshaw, Paul. "Breath of Death." *Monsanto Magazine* 25 (Dec. 1946): 26–27.

Herken, Gregg. "The Earthly Origins of Star Wars." *Bulletin of the Atomic Scientists* 43 (Oct. 1987): 20–28.

Hirst, Eric. "Transportation Energy Use and Conservation Potential." *Bulletin of the Atomic Scientists* 29 (Nov. 1973): 36–42.

Hirst, Eric, and Janet Carney. "Effects of Federal Residential Energy Conservation Programs." *Science* 199 (24 Feb. 1978): 845–51.

"Hive Technology." *Scientific American* 260 (Jan. 1989): 22.

Holdren, John. "Harrison Brown, 1917–1986." *Bulletin of the Atomic Scientists* 43 (Mar. 1987): 3–8.

Hollaender, Alexander. "Modification of Radiation Response." *Bulletin of the Atomic Scientists* 12 (Mar. 1956): 76–80.

Holton, Gerald. "The Migration of Physicists to the United States." *Bulletin of the Atomic Scientists* 40 (Apr. 1984): 18–24.

Jefferson, Jon. "Report from Atom City." *Newsweek*, 7 Jan. 1991, 9.

Johnson, Warren, Laurence Quill, and Farrington Daniels. "Rare Earths Separation Developed on Manhattan Project." *Chemical and Engineering News* 25 (1 Sept. 1947): 2494.

Kerr, Richard A. "Indoor Radon: The Deadliest Pollutant." *Science* 240 (29 Apr. 1988): 606–8.

"Labs Struggle to Promote Spin-Offs." *Science* 240 (13 May 1988): 874–75.

Lacassagne, A. "The Risks of Cancer Formation by Radiations." *Bulletin of the Atomic Scientists* 12 (Apr. 1957): 135–42.

Lanouette, William. "Bumbling toward the Bomb." *Bulletin of the Atomic Scientists* 45 (Sept. 1989): 7–11.

Lapp, Ralph. "Sunshine and Darkness." *Bulletin of the Atomic Scientists* 15 (Jan. 1959): 27–29.

Lay, Fernando. "Nuclear Technology in Outer Space." *Bulletin of the Atomic Scientists* 35 (Sept. 1979): 27–31.

Lear, John. "Spinning the Thread of Life." *Saturday Review*, 5 Apr. 1969, 63–66.

LeRoy, George. "Atomic Energy and the Life Sciences." *Bulletin of the Atomic Scientists* 5 (Nov. 1949): 315–22.

Leverett, Miles. "Highlights of Reactor Technology, 1942–1955." *General Electric Review*, Nov. 1955, 23–26.

Lewis, Richard. "The Radioactive Salt Mine." *Bulletin of the Atomic Scientists* 27 (June 1971): 27–33.

Lind, Samuel. "The Memoirs of Samuel Colville Lind." *Journal of the Tennessee Academy of Science* 47 (Jan. 1972): 1–40.

Livingston, Robert. "Acceleration of Partially Stripped Heavy Ions." *Nature* 173 (9 Jan. 1954): 54–55.

———. "Samuel Colville Lind: The First Eighty Years." *Radiation Research* 10 (June 1959): 604–6.

Love, Leon. "Electromagnetic Separation of Isotopes at Oak Ridge." *Science* 182 (26 Oct. 1973): 343–52.

Lovins, Amory. "The Case against the Fast Breeder Reactor." *Bulletin of the Atomic Scientists* 29 (Mar. 1973): 29–35.

Lum, James. "Baker Day." *Monsanto Magazine* 25 (Dec. 1946): 25.

Manhattan Project Headquarters. "Availability of Radioactive Isotopes." *Science* 103 (14 June 1946): 697–705.

Manley, John. "Assembling the Wartime Labs." *Bulletin of the Atomic Scientists* 30 (May 1974): 42–48.

McCullough, C. Rogers. "Harnessing the Atom." *Monsanto Magazine* 25 (Dec. 1946): 28–29.

Meier, Richard. "The Social Impact of a Nuplex." *Bulletin of the Atomic Scientists* 25 (Mar. 1969): 16–19.

Micklin, Philip. "Environmental Hazards of Nuclear Wastes." *Bulletin of the Atomic Scientists* 30 (Apr. 1974): 36–42.

"MIT Tokamak Alcator C Exceeds Lawson Criterion." *Physics Today* 37 (Feb. 1984): 20–22.

Morgan, Karl. "Shipping of Radio-Isotopes." *Journal of Applied Physics* 19 (July 1948): 593–98.

———. "Human Exposure to Radiation." *Bulletin of the Atomic Scientists* 15 (Nov. 1959): 384–89.

Morris, Gibson. "Tower for Radiation Testing Meets Unusual Requirements." *Civil Engineering* 26 (Sept. 1956): 68–70.

Nash, William, and O. L. Persechini, "The First 1000 Days." *Monsanto Magazine* 25 (Feb. 1946): 4–11.

Neir, Alfred. "The Mass Spectrometer." *Scientific American* 188 (Mar. 1953): 68–74.

"Neutron Scattering: A New National Facility at Oak Ridge." *Science* 199 (10 Feb. 1978): 673.

Newman, Stanley. "Civil Defense and the Congress: Quiet Reversal." *Bulletin of the Atomic Scientists* 18 (Nov. 1962): 33–37.

Norris, Dewey, and Arthur Snell. "Production of C14 in a Nuclear Reactor." *Nucleonics* 5 (Sept. 1949): 18–27.

"Nuclear Research Institute at Oak Ridge." *Science* 103 (14 June 1946): 705–6.

ORMAK-ISX Group. "Results of the ORMAK, ISX-A, and ISX-B Programmes." *Nuclear Fusion* 25 (Sept. 1985): 1137–43.

Parsegian, V. Lawrence. "National Laboratories Versus National Progress." *Rensselaer Alumni News*, June 1962, 2–5.

———. "Makework Projects Waste U.S. Brain Power." *Nation's Business*, Sept. 1962, 106–7.

———. "On the Role of Government Laboratories." *Bulletin of the Atomic Scientists* 22 (Sept. 1966): 35–36.

Pauling, Linus. "Genetic Effects of Weapons Tests." *Bulletin of the Atomic Scientists* 18 (Dec. 1962): 15–18.

Powers, Philip. "The History of Nuclear Engineering Education." *Journal of Engineering Education* 54 (June 1964): 364–70.

"Report of President Truman on the Atomic Bomb." *Science* 102 (17 Aug. 1945): 163–65.

Richardson, Robert. "The Selling of the Atom." *Bulletin of the Atomic Scientists* 30 (Oct. 1974): 28–35.

Rose, David. "Nuclear Power." *Bulletin of the Atomic Scientists* 41 (Aug. 1985): 76–78.

Ross, Douglas. "Thermonuclear's Stepchild." *Oak Ridge National Laboratory Review* 2 (Fall 1968): 14–19.

"Roundtable: New Challenges for the National Labs." *Physics Today* 44 (Feb. 1991): 24–35.

Rush, Joseph. "Prometheus in Tennessee." *Saturday Review*, 2 July 1960, 10–11, 50.

Rudolph, Philip. "Samuel Colville Lind's Illustrious Career Matched Modern Science Development." *Oak Ridge National Laboratory News*, June 18, 1965.

Samios, N. P. "Brookhaven and SDI." *Bulletin of the Atomic Scientists* 42 (Mar. 1986): 56–57.

Seitz, Frederick, and Eugene Wigner. "The Effects of Radiation on Solids." *Scientific American* 195 (Aug. 1956): 76–84.

Selove, Walter, and Mortimer Elkind. "Radiation and Man." *Bulletin of the Atomic Scientists* 14 (Jan. 1958): 7–8.

Shils, Edward. "Freedom and Influence: Observations on the Scientists" Movement in the United States." *Bulletin of the Atomic Scientists* 12 (Jan. 1957): 13–18.

Smith, Alice. "Behind the Decision to Use the Atomic Bomb: Chicago, 1944–45." *Bulletin of the Atomic Scientists* 14 (Oct. 1958): 288–312.

———. "Scientists and Public Issues." *Bulletin of the Atomic Scientists* 38 (Dec. 1982): 38–45.

Smyth, Henry. "The Role of the National Laboratories in Atomic Energy Development." *Bulletin of the Atomic Scientists* 6 (Jan. 1950): 5–8.

Snell, Arthur. "Exciting Years!" *Nuclear Science and Engineering* 90 (1985): 358–66.

Snell, Arthur, Frances Pleasonton, and R. V. McCord, "Radioactive Decay of the Neutron." *Physical Review* 78 (1 May 1950): 310–11.

Snell, Arthur, and Alvin Weinberg. "History and Accomplishments of the Oak Ridge Graphite Reactor." *Physics Today* 17 (Aug. 1964): 32–38.

Snell, Arthur, et al. "The Cyclotron Group at the Metallurgical Laboratory, Chicago, 1940–44." *American Journal of Physics* 48 (Nov. 1980): 971–78.

Spinrad, Bernard. "Why Not National Laboratories?" *Bulletin of the Atomic Scientists* 22 (Apr. 1966): 20–23.

Stead, William. "The Sun and Foreign Policy." *Bulletin of the Atomic Scientists* 12 (Mar. 1957): 86–90.

Steiner, Arthur. "Baptism of the Atomic Scientists." *Bulletin of the Atomic Scientists* 31 (Feb. 1975): 21–28.

"Still Going Strong in Tennessee." *Nature* 220 (16 Nov. 1968): 640–41.

Strauss, Lewis. "The U.S. Atomic Energy Program, 1953–58." *Bulletin of the Atomic Scientists* 14 (Sept. 1958): 256–58.

"Strontium-90—Some Notes on Present and Future Levels." *Bulletin of the Atomic Scientists* 14 (Jan. 1958): 62.

Struxness, Edward, and Roy Morton. "Ground Disposal of Radioactive Wastes." *American Journal of Public Health* 46 (1956): 56–63.

Stubbs, James, and Latresia Wilson. "Nuclear Medicine: A State-of-the-Art Review." *Nuclear News* 34 (May 1991): 50–54.

Svirsky, Leon. "The Atomic Energy Commission." *Scientific American* 181 (July 1949): 30–43.

Szilard, Leo. "Leo Szilard: His Version of the Facts." *Bulletin of the Atomic Scientists* 35 (Mar.–Apr. 1979): 28–32, 55–59.

Taylor, Ellison. "Samuel Colville Lind." *Journal of Physical Chemistry* 63 (1959): 772–76.

Teller, Edward. "John von Neumann." *Bulletin of the Atomic Scientists* 12 (Apr. 1957): 150–51.

———. "The Era of Big Science." *Bulletin of the Atomic Scientists* 27 (Apr. 1971): 34–36.

Thomas, David, Phil Hayes, William Mixon, John Sheppard, William Griffith, and Robert Keller. "Turbulence Promoters for Hyperfiltration with Dynamic Membranes." *Environmental Science and Technology* 4 (Dec. 1970): 1129–36.

Thompson, Dick. "Living Happily Near a Nuclear Trash Heap." *Time,* 11 May 1992, 53–54.

Tierney, John. "Take the A-Plane." *Science* 82 (3 Jan. 1982): 46–55.

Tompkins, Paul. "Radioisotopes." *Monsanto Magazine* 25 (Dec. 1946): 32–34.

Trivelpiece, Alvin. "Competitiveness Begins at Home." *Oak Ridge National Laboratory Review* 22, no. 1 (1989): 22–27.

Walsh, John. "Oak Ridge: 20 Years After, Diversification IS the Goal." *Science* 150 (12 Nov. 1965): 863–65.

———. "Oak Ridge National Laboratory: Aim Is Change along with Growth." *Science* 150 (26 Nov. 1965): 1133–36.

———. "A Conversation with Eugene Wigner." *Science* 181 (10 Aug. 1973): 527–33.

———. "DOE Laboratories in the Spotlight." *Science* 213 (14 Aug. 1981): 744.

———. "Commerce to Inherit Energy Research." *Science* 215 (8 Jan. 1982): 147–48.

Wattenberg, Albert. "The Building of the First Chain Reaction Pile." *Bulletin of the Atomic Scientists* 30 (June 1974): 51–57.

Weinberg, Alvin. "Oak Ridge National Laboratory." *Science* 109 (11 Mar. 1949): 245–48.

———. "How Shall We Establish a Nuclear Power Industry in the United States?" *Bulletin of the Atomic Scientists* 9 (May 1953): 120–23.

———. "Future Aims of Large Scale Research." *Chemical and Engineering News* 47 (23 May 1955): 2188–91.

———. "Today's Revolution." *Bulletin of the Atomic Scientists* 12 (Oct. 1956): 299–302.

———. "Prospects in International Science." *Bulletin of the Atomic Scientists* 14 (Dec. 1958): 402–4.

———. "Energy as an Ultimate Raw Material, or Problems of Burning the Sea and Burning the Rocks." *Physics Today* 12 (Nov. 1959): 12.

———. "The Federal Laboratories and Science Education." *Science* 136 (6 Apr. 1962): 27–29.

———. "Samuel K. Allison, 1901–1965." *Bulletin of the Atomic Scientists* 22 (Jan. 1966): 2.

———. "Science, Choice, and Human Values." *Bulletin of the Atomic Scientists* 22 (Apr. 1966): 8–13.

———. "Can Technology Replace Social Engineering?" *Bulletin of the Atomic Scientists* 22 (Dec. 1966): 4–8.

———. "In Defense of Science." *Science* 167 (9 Jan. 1970): 141–45.

———. "Social Institutions and Nuclear Energy." *Science* 177 (7 July 1972): 27–34.

———. "The Many Dimensions of Scientific Responsibility." *Bulletin of the Atomic Scientists* 32 (Nov. 1976): 21–29.

———. "Is Nuclear Energy Acceptable?" *Bulletin of the Atomic Scientists* 33 (Apr. 1977): 54–60.

———. "Is Nuclear Energy Necessary?" *Bulletin of the Atomic Scientists* 36 (Mar. 1980): 31–35.

———. "Avoiding the Entropy Trap." *Bulletin of the Atomic Scientists* 38 (Oct. 1982): 32–35, 61–62.

———. "A Nuclear Power Advocate Reflects on Chernobyl." *Bulletin of the Atomic Scientists* 42 (Aug. 1986): 57–60.

Weinberg, Alvin, Jack Barkenbus, and Michael Alonso. "Storing the World's Spent Nuclear Fuel." *Bulletin of the Atomic Scientists* 41 (Nov. 1985): 34–37.

Weinberg, Alvin, C. C. Burwell, and M. J. Ohanian, "A Siting Policy for an Acceptable Nuclear Future." *Science* 204 (June 1979): 1043–51.

Weinberg, Alvin, and Lawrence Dresner. "Recent Developments in the Theory and Technology of Chain Reactors." *Reviews of Modern Physics* 34 (Oct. 1962): 747–65.

Weinberg, Alvin, and R. Philip Hammond. "Limits to the Use of Energy." *American Scientist* 58 (July 1970): 412–18.

———. "Global Effects of Increased Use of Energy." *Bulletin of the Atomic Scientists* 28 (Apr. 1972): 5–8.

Weisner, Jerome. "On Science Advice to the President." *Scientific American* 260 (Jan. 1989): 34–39.

Weisskopf, Victor. "Why Pure Science?" *Bulletin of the Atomic Scientists* 21 (Apr. 1965): 4–8.

Whitaker, Martin. "Dogpatch." *Monsanto Magazine* 24 (Dec. 1945): 14–17.

Wigner, Eugene. "Theoretical Physics in the Metallurgical Laboratory of Chicago." *Journal of Applied Physics* 17 (Nov. 1946): 857–63.

———. "Atomic Energy." *Science* 108 (12 Nov. 1948): 517–21.

———. "Fallout: Criticism of a Criticism." *Bulletin of the Atomic Scientists* 16 (Mar. 1960): 107–8.

———. "Recall the Ends—While Pondering Means." *Bulletin of the Atomic Scientists* 17 (Mar. 1961): 82–85.

———. "Why Civil Defense?" *Technology Review* 66 (June 1964): 1–4.

———. "Violations of Symmetry in Physics." *Scientific American* 213 (Dec. 1965): 28–36.

Wigner, Eugene, and Frederick Seitz. "Pure and Applied Nuclear Physics in East and West." *Bulletin of the Atomic Scientists* 15 (Mar. 1959): 127–37.

Wigner, Eugene, and Alvin Weinberg. "Longer Range View of Nuclear Energy." *Bulletin of the Atomic Scientists* 16 (Dec. 1960): 400–403.

Wilkinson, Michael. "Early History of Neutron Scattering at Oak Ridge." *Physica* 137B (1986): 3–16.

Winkler, Allan. "A 40-Year History of Civil Defense." *Bulletin of the Atomic Scientists* 40 (June 1984): 16–22.

Wollan, Ernest. "The Other Record of the First Nuclear Reactor Start-up." *American Journal of Physics* 48 (Nov. 1980): 979–80.

Young, Gale. "The Coming of Age of Atomic Power." *Nuclear News* 6 (Dec. 1963): 6–12.

———. "The Fueling of Nuclear Power Complexes." *Nuclear News* 7 (Nov. 1964): 23–30.

Young, Gale, and Alvin Weinberg. "Scale, Nuclear Economics and Salt Water." *Nuclear News* 6 (May 1963): 3.

Zucker, Alexander. "State of the Laboratory, 1978–88: Years of Change." *Oak Ridge National Laboratory Review* 22, no. 1 (1989): 2–21.

Zucker, Alexander, Bernard Cohen, and H. L. Reynolds, "A Comparison of Nitrogen- and Proton-Induced Nuclear Reactions." *Physical Review* 96 (15 Dec. 1954): 1617.

Zucker, Alexander, and L. D. Wyly, "Activities in Light Nuclei from Nitrogen Ion Bombardment." *Physical Review* 89 (15 Jan. 1953): 524.

Books

Ahnfeldt, Arnold, ed., *Radiology in World War II*. Washington, D.C.: Office of the Surgeon General, 1966.

Allardice, Corbin. *The Atomic Energy Commission*. New York: Praeger Publishers, 1974.

Allardice, Corbin, and Edward Trapnell. *The First Pile*. Oak Ridge: Atomic Energy Commission, 1949.

Bishop, Amasa. *Project Sherwood: The U.S. Program of Controlled Fusion*. Reading, Mass.: Addison-Wesley Publishing Co., 1958.

Brown, Harrison, James Bonner, and John Weir. *The Next Hundred Years*. New York: Viking Press, 1954.

Bupp, Irvin, and Jean-Claude Derian. *Light Water: How the Nuclear Dream Dissolved*. New York: Basic Books, 1978.

Charpie, Robert, J. Horowitz, D. J. Hughes, and D. J. Littler, *Physics and Mathematics*. New York: McGraw-Hill Book Co., 1956.

Clawson, Marion, and Hans Landsberg. *Desalting Seawater: Achievements and Prospects*. New York: Gordon and Breach Science Publishers, 1972.

Cohen, I. Bernard. *Revolution in Science*. Cambridge, Mass.: Harvard Univ. Press, 1985.

Coryell, Charles, ed. *Radiochemical Studies: The Fission Products*. New York: McGraw-Hill Book Co., 1951.

Dean, Prentice. *Energy History Chronology from World War II to the Present*. Washington, D.C.: Department of Energy DOE/ES-0002, 1982.

Gailar, Joanne. *Oak Ridge and Me, From Youth to Maturity*. Oak Ridge: Children's Museum, 1991.

Glasstone, Samuel. *Sourcebook on Atomic Energy*. New York: D. Van Nostrand Co., 1950.

———. *Principles of Nuclear Reactor Engineering*. New York: D. Van Nostrand Co., 1955.

Glasstone, Samuel, and M. C. Edlund, *The Elements of Nuclear Reactor Theory*. New York: D. Van Nostrand Co., 1952.

Greenbaum, Leonard. *A Special Interest: The Atomic Energy Commission, Argonne National Laboratory, and the Midwestern Universities*. Ann Arbor: Univ. of Michigan Press, 1971.

Groueff, Stephane. *Manhattan Project: The Untold Story of the Making of the Atomic Bomb*. Boston: Little, Brown and Company, 1967.

Groves, Leslie. *Now It Can Be Told: The Story of the Manhattan Project*. New York: Harper & Bros., 1962.

Hacker, Barton. *The Dragon's Tail: Radiation Safety in the Manhattan Project, 1942–1946.* Berkeley: Univ. of California Press, 1987.

Heilbron, J. L., and Robert Seidel. *Lawrence and His Laboratory: A History of the Lawrence Berkeley Laboratory.* Vol. 1. Berkeley: Univ. of California Press, 1989.

Hewlett, Richard. *Nuclear Navy, 1946–1962.* Chicago: Univ. of Chicago Press, 1974.

Hewlett, Richard, and Oscar Anderson, Jr. *A History of the United States Atomic Energy Commission.* Vol. 1, *The New World, 1939–1946.* Washington, D.C.: Atomic Energy Commission, 1972.

Hewlett, Richard, and Francis Duncan. *A History of the Unites States Atomic Energy Commission.* Vol. 2, *Atomic Shield, 1947–1952.* University Park: Univ. of Pennsylvania Press, 1969.

Hewlett, Richard, and Jack Holl. *Atoms for Peace and War, 1953–1961.* Berkeley: Univ. of California Press, 1989.

Hirshleifer, Jack, C. De Haven, and Jerome Milliman. *Water Supply: Economics, Technology, and Policy.* Chicago: Univ. of Chicago Press, 1960.

History Associates, Inc. *The Federal Role and Activities in Energy Research and Development, 1946–1980: An Historical Summary.* Oak Ridge: Oak Ridge National Laboratory for Department of Energy, 1983.

Holl, Jack, Roger Anders, and Alice Buck. *United States Civilian Nuclear Power Policy, 1954–1984: A Summary History.* Washington, D.C.: Department of Energy, 1986.

Jones, Vincent. *Manhattan: The Army and the Atomic Bomb.* United States Army in World War II series. Washington, D.C.: Center of Military History, 1985.

Kevles, Daniel. *The Physicists: The History of a Scientific Community in Modern America.* Cambridge, Mass.: Harvard Univ. Press, 1987.

Kohlstedt, Sally G., and Margaret W. Rossiter. *Historical Writing on American Science.* Baltimore: Johns Hopkins Univ. Press, 1985.

Libby, Leona. *The Uranium People.* New York: Charles Scribner's Sons, 1979.

Lilienthal, David. *The Journals of David E. Lilienthal.* 2 vols. New York: Harper and Row, 1964.

Lustman, Benjamin, and Frank Kerze, eds. *The Metallurgy of Zirconium.* New York: McGraw-Hill Book Co., 1955.

Lyon, R. N., R. Hurst, and C. M. Nicholls, *Technology, Engineering, and Safety.* New York: Pergamon Press, 1960.

Moon, Ralph. *Proceedings of the 3rd International Conference on Neutron Scattering.* Gatlinburg, Tennessee, June 6–10, 1976. Oak Ridge: Oak Ridge National Laboratory CONF-760601-P1, 1976.

National Academy of Engineering. *Priorities for Research Applicable to National Needs.* Washington, D.C.: National Academy of Engineering, 1973.

Nelson, Daniel, ed. *Radionuclides in Ecosystems: Proceedings of the Third National Symposium on Radioecology.* Washington, D.C.: AEC Report CONF-710501, 1973.

Nichols, Kenneth. *The Road to Trinity.* New York: William Morrow and Co., 1987.

Office of Technology Assessment. *Complex Cleanup: The Environmental Legacy of Nuclear Weapons Production.* Washington, D.C.: Government Printing Office, 1991.

Overholt, James, ed. *These Are Our Voices: The Story of Oak Ridge, 1942–1970*. Oak Ridge: Children's Museum, 1987.

Pilat, Joseph, Robert Pendley, and Charles Ebinger, eds. *Atoms for Peace: An Analysis after Thirty Years*. Boulder, Colo.: Westview Press, 1985.

Pollard, William. *ORAU: From the Beginning*. Oak Ridge: Oak Ridge Associated Universities, 1980.

Polmar, Norman, and Thomas Allen. *Rickover*. New York: Simon and Schuster, 1982.

Post, Roy, and Robert Seale, eds. *Water Production Using Nuclear Energy*. Tucson: Univ. of Arizona Press, 1966.

Rhodes, Richard. *The Making of the Atomic Bomb*. New York: Simon and Schuster, 1986.

Russell, Milton, Bruce Tonn, Ho-Ling Hwang, Richard Goeltz, and John Warren. *Hazardous Waste Remediation Project: Cost of RCRA Corrective Action*. Knoxville: Waste Management Research and Education Institute, 1992.

Sachs, Robert, ed. *The Nuclear Chain Reaction—Forty Years Later*. Chicago: Univ. of Chicago Press, 1984.

Sanger, S. L., and Robert W. Mull. *Hanford and the Bomb: An Oral History of World War II*. Seattle: Living History Press, 1989.

Seaborg, Glenn, Joseph Katz, and Winston Manning, eds. *The Transuranium Elements Research Papers*. New York: McGraw-Hill Book Co., 1949.

Smyth, Henry. *Atomic Energy for Military Purposes*. Princeton: Princeton Univ. Press, 1945.

Spiegler, K. S., ed. *Principles of Desalination*. New York: Academic Press, 1966.

Stannard, J. Newell. *Radioactivity and Health: A History*. Washington, D.C.: Office of Scientific and Technical Information, 1988.

Strauss, Lewis. *Men and Decisions*. London: Macmillan and Co., 1963.

Strickland, Donald. *Scientists in Politics: The Atomic Scientists Movement, 1945–1946*. Lafayette, Ind.: Purdue Univ. Studies, 1968.

Sweet, Colin. *The Fast Breeder Reactor: Need? Cost? Risk?* London: Macmillan Press, 1980.

Symposium on the Use of Isotopes in Biology and Medicine. Madison: Univ. of Wisconsin Press, 1948.

Sylves, Richard. *The Nuclear Oracles: A Political History of the General Advisory Committee of the Atomic Energy Commission, 1947–1977*. Ames: Iowa State Univ. Press, 1987.

Taub, A. H., ed. *John von Neumann, Collected Works*. 6 vols. New York: Pergamon Press, 1961–63.

The World Nuclear Handbook. London: Euromonitor Publications, 1988.

Weart, Spencer. *Scientists in Power*. Cambridge, Mass.: Harvard Univ. Press, 1979.

Weinberg, Alvin. *Reflections on Big Science*. Cambridge, Mass.: Massachusetts Institute of Technology Press, 1967.

Weinberg, Alvin, and Eugene Wigner. *The Physical Theory of Neutron Chain Reactors*. Chicago: Univ. of Chicago Press, 1958.

Whicker, F. Ward, and Vincent Schultz. *Radioecology: Nuclear Energy and the Environment*. 2 vols. Boca Raton, Fla.: CRC Press, 1982.

Wigner, Eugene. *Symmetries and Reflections*. Bloomington: Indiana Univ. Press, 1967.

———, ed. *Survival and the Bomb: Methods of Civil Defense*. Bloomington: Indiana Univ. Press, 1969.

Zirkle, Raymond, ed. *Biological Effects of External X and Gamma Radiation*. New York: McGraw-Hill Book Co., 1954.

Laboratory/Departmental Publications

Dean, Prentice. *Energy History Chronology from World War II to the Present*. Washington, D.C.: DOE/ER-0002, 1982.

Greene, Harold, and Margie Skipper. *History of the Laboratory Protection Division, Oak Ridge National Laboratory, 1942–1992*. Oak Ridge: Laboratory Protection Division, 1992.

Oak Ridge National Laboratory. *High Voltage Laboratory*. Oak Ridge: Oak Ridge National Laboratory, 1958.

———. *High Voltage Accelerators at Oak Ridge National Laboratory*. Oak Ridge: Oak Ridge National Laboratory, n.d.

———. *Cyclotrons at Oak Ridge National Laboratory*. Oak Ridge: Oak Ridge National Laboratory, n.d.

———. *The Bulk Shielding Reactor*. Oak Ridge: Oak Ridge National Laboratory, 1961.

———. *ORNL Directions, 1979–1984*. Oak Ridge: Oak Ridge National Laboratory, 1979.

———. *ORNL Trends and Balances, 1984–1989*. Oak Ridge: Oak Ridge National Laboratory, 1984.

———. *ORNL '90*. Oak Ridge: Oak Ridge National Laboratory, 1990.

———. *A Climate for Collaboration: Oak Ridge National Laboratory's 1990 R&D 100 Award Winners*. Oak Ridge: Oak Ridge National Laboratory, 1991.

U.S. Atomic Energy Commission. *The Future Role of the Atomic Energy Commission Laboratories*. Washington, D.C.: Atomic Energy Commission, 1960.

———. *Annual Reports*. Washington, D.C.: Atomic Energy Commission, 1946–1973.

U.S. Department of Energy. Office of Energy Research. *Capsule Review of DOE Research and Development and Field Facilities*. Washington, D.C.: DOE/ER-0305, 1986.

———. *The Greenhouse Effect*. Washington, D.C.: DOE/ER-0411, 1989.

Documents

U.S. Comptroller General. *The Multiprogram Laboratories: A National Resource for Non-nuclear Energy Research, Development, and Demonstration*. Washington, D.C.: Comptroller General Report EMD-78-62, 1978.

U.S. Congress. Joint Committee on Atomic Energy. *Aircraft Nuclear Propulsion Program*. 86th Cong., 1st sess., 1959. Committee Print.

———. *The Future Role of the Atomic Energy Commission Laboratories*. 86th Cong., 2d sess., 1960. Committee Print.

————. *To Designate the Oak Ridge National Laboratory, Oak Ridge, Tennessee, as the "Holifield National Laboratory."* H. Report No. 93-1612, 93d Cong., 2d sess., 1974.

U.S. Congress. House. Committee on Science and Astronautics. *Towards a Science Policy for the United States.* 91st Cong., 2d sess., 1970. Committee Print.

————. Committee on Science and Technology. *The Role of the National Energy Laboratories in ERDA and Department of Energy Operations: Retrospect and Prospect.* Committee Report by Library of Congress Science Policy Research Division. 95th Cong., 2d sess., 1978.

————. *Carbon Dioxide and Climate: The Greenhouse Effect.* 97th Cong., 1st sess., 1981. Committee Print 45.

————. *Department of Energy Authorization.* 97th Cong., 1st sess., 1981. Committee Print 37.

————. *Department of Energy, Fiscal Year 1983, Budget on Environment, Health, and Safety.* 97th Cong., 2d sess., 1982. Committee Print 97.

————. *Fiscal Year 1986 DOE Budget Authorization: Environmental Research and Development.* 99th Cong., 1st sess., 1985. Committee Print 38.

U.S. Congress. Office of Technology Assessment. *Complex Cleanup: The Environmental Legacy of Nuclear Weapons Production.* Washington, D.C.: Government Printing Office, 1991.

U.S. Congress. Senate. *Report of the United States Atomic Energy Commission.* Sen. Doc. No. 118, 80th Cong., 2d sess., 1948.

————. Committee on Public Works. *A Case for National Environmental Laboratories.* 91st Cong., 2d sess., 1970. Committee Print.

Manuscripts

Abner, C. H., V. T. Carmody, J. R. Hensley, J. J. Varagona, G. B. Young, and Peggy Geldmeier. "History of the Plant and Equipment Division." Oak Ridge, 1992.

Auerbach, Stanley I. "A History of the Environmental Sciences Division of Oak Ridge National Laboratory." Oak Ridge, 1992.

n.a. "A Brief History of the Research Reactors Division of Oak Ridge National Laboratory." Oak Ridge, 1992.

Childress, C. E., R. E. Fenstermaker, and Amy L. Harkey. "A Brief History: The Quality Department of Oak Ridge National Laboratory, 1955–1991." Oak Ridge, 1991.

n.a. "Finance and Business Management Division History." Oak Ridge, 1992.

Greenstreet, William L. "Draft: Abridged History of the Engineering Technology Division." Oak Ridge, 1992.

Kerr, G. D., R. H. Ritchie, S. V. Kaye, and J. S. Wassom. "A Brief History of the Health and Safety Research Division." Oak Ridge, 1992.

Love, Leon. "The Early History of the Electromagnetic Separation of Large Quantities of Stable and Radioactive Isotopes." Oak Ridge, 1991.

n.a. "The ORNL Fusion Energy Division History." Oak Ridge, 1992.

Thompson, William. "History of the Oak Ridge National Laboratory, 1943–1963." Oak
    Ridge: ORNL Central Files 63-8-75, 1963.

Watkins, Michael. "The History of Radiation Protection at Oak Ridge National Labora-
    tory." Oak Ridge, 1992.

Weinberg, Alvin. "Autobiography." Draft by Alvin Weinberg. Oak Ridge, 1992.

Wilkinson, Michael. "ORNL Solid State Division, 1952–1992." Oak Ridge, 1992.

Zimmerman, K. H. "History of the Energy Division." Oak Ridge, 1992.

News Sources

Knoxville, Tenn. *Knoxville News-Sentinel*.

New York, N.Y. *New York Times*.

Oak Ridge, Tenn. *Energy Systems News*. 1984–92.

———. *Nuclear Division News*. 1970–84.

———. *Oak Ridge National Laboratory News*. 1948–70.

———. *Oak Ridge National Laboratory Review*. 1967–92.

———. *The Oak Ridger*.

Archives

Knoxville, Tenn. Univ. of Tennessee Library. Special Collections. Alexander Hollaender
    Papers.

———. Oak Ridge National Laboratory Biology Division Papers.

———. Richard Setlow Papers.

Oak Ridge, Tenn. Children's Museum. Alvin M. Weinberg Papers.

———. Oak Ridge National Laboratory. Director's Files.

———. Oak Ridge National Laboratory. Central Files.

———. Oak Ridge National Laboratory. Historical Archives.

Index

Greenbaum, Eli, 174
Griffith, William, 164
Grimes, Warren, 44–45
Groves, Leslie, 17, 28, 30, 46, 48, 58
Guerin, Mike, 231
Gulf General Atomic Corporation, 135

Hamel, William, 201
Hamilton, Joseph, 140
Hammond, Philip, 110, 111
Hancher, Charles, 164, 197
Hanford, Wash., plutonium production in, 24–26
Happer, William, 220
Harms, William, 135
Harrell, William, 50
Hartman, Fred, 193
Haubenreich, Paul, 170, 220–21
Haynes, Virgil, 165
Hayter, John, 211, 235
health physics research reactor, 134–35, 213
Heavy Section Steel Program, 123–24, 147
Hendricks, Robert, 182
Hermann, Grover, 189–90
Herndon, Joseph, 201
heterogeneous reactors, 81
Hibbs, Roger, 189
high-flux reactor, 130–33, 213
High Temperature Materials Laboratory, 196, 198–99, 233
Hilberry, Norman, 11
Hirst, Eric, 150, 165
Hise, Eugene, 167
Hobson, David, 148, 177
Holifield, Chet, 124–25, 141, 158–59, 180
Holifield Heavy Ion Research Facility, 141, 159, 171, 179, 180–81, 196, 239, 240
Holifield National Laboratory, 158–59
Hollaender, Alexander, 31, 37–38, 107, 112

Holland, Leo, 80
Holman, Alan, 167
Holsopple, Herman, 114
Holt, Andrew, 115
homogeneous reactors, 80–83, 97–98
Honea, Robert, 167
Hood Laboratory, 92
Hoskins, Robert, 165
Householder, Alston, 70, 72
Housing and Urban Development (HUD), Department of, 162
HUD. *See* Housing and Urban Development, Department of
Human Genome Management Information System, 225
Human Genome Project, 194
human genome research, 194, 224–25
Hurst, Samuel, 176
hydrocarbon reactor, 166
hydrogen plasma, 142–43

Idaho National Engineering Laboratory, 57, 58, 235
impurities studies experiments (ISXs), 169
indoor air pollution, 192
information centers, 107–9
INOR-8, 61–62
International Fusion Superconducting Magnet Test Facility, 219
International Nickel Company, 62
International Thermonuclear Energy Reactor (ITER), 220–21
ISXs. *See* impurities studies experiments

Jackson, Bo, 194
Jared, Don, 230
Johnson, Cleland, 86
Johnson, Dale, 217
Johnson, Lyndon, 110
Johnson, Warren, 50
Johnson, William, 37
Johnson and Johnson Corporation, 194

nuclear fuel: purification of, 84–86; reprocessing of, 33
nuclear piles, first, 11–15
nuclear power, peacetime applications
of, 53–54, 78–80
nuclear-powered aircraft, 58–63, 76–77,
96
nuclear-powered ship, 77
nuclear reactors: advanced neutron
source (ANS), 234–36; cooling of, 15,
60; gas-cooled reactors, 89, 134, 135;
graphite reactor, 11–15, 20, 21, 22–23;
health physics research reactor, 134–
35, 213; high-flux reactor, 130–33, 213;
homogeneous reactors, 80–83, 97–98;
liquid-metal fast breeder reactor, 135–
36 materials testing reactor, 32–33, 57;
molten-salt reactor, 90, 136–38, 154;
package reactor, 83–84; research budget, 202–3; research reactor, 86
Nuclear Regulatory Commission (NRC),
128, 155, 160
nuclear safety, 123–25, 147–49, 170, 202–
3, 208; and Three Mile Island, 176–78
nuclear wastes, 143–44, 192, 226–28

Oak Ridge, Tenn.: geographic setting of,
1–2
Oak Ridge Associated Universities, 40,
153, 181
Oak Ridge Automatic Computer and
Logical Engine. See ORACLE com-
puter
Oak Ridge Detector Center, 211
Oak Ridge Electron Linear Accelerator
(ORELA), 138–39
Oak Ridge Institute of Nuclear Studies,
40, 49, 51
Oak Ridge Isochronous Cyclotron
(ORIC), 138, 141
Oak Ridge National Laboratory
(ORNL), 3–5, 28–29, 52; Analytical
Chemistry Division, 85, 114, 230;
Biology Division, 37–39, 112–15, 159,
193, 194, 224, 225; budget cuts at, 96–
97, 126–27, 151–52, 185; and Jimmy
Carter, 173, 175–76; Center for Computational Sciences, 211–12; Center
for Global Studies, 211; Chemical
Technology Division, 84, 85, 101, 164,
174, 177, 193, 197; Chemistry Division, 173, 174; civil defense efforts,
115–18; Clinch River breeder reactor
project, 170, 178–79; competition
faced by, 103–5; Defense Department
work 229–30; and the Department of
Energy, 172–75; desalination efforts,
108–12; education efforts, 39–40, 238;
Electronuclear Division, 65, 69, 83,
92–93, 138; and energy crisis, 149–51,
154–55, 159–61; Energy Division, 124,
161, 174, 237; Engineering Technology Division, 170, 174, 177; Environmental, Life, and Social Sciences
Complex, 239–40; Environmental
Sciences Division, 100, 124, 125, 146,
173, 192, 216, 217–18; and ERDA,
160–61; expansion of, 171–72, 179–82;
future challenges, 234–36, 237–42; as
global science center, 205; graphite
reactor at, 20, 21, 22–23; Health and
Safety Research Division, 192, 193,
194, 195, 225; Health Physics Division, 99, 100, 101, 119; Industrial
Safety and Applied Health Physics
Division, 177; Instrumentation and
Controls Division, 120, 177; Isotopes
Division, 119; long-term strategy of,
237–41; management of, 46–49, 191,
205–9; Martin Marietta Energy Systems as manager of, 189–91; Mathematics and Computing Section, 70–
72; Metallurgy Division, 37, 44;
Metals and Ceramics Division, 131,
171, 173–74, 214; Monsanto's management of, 29–32, 48–49; Neutron